Praise Your Way Through

AF808060

Iris Celeste

This book is a work of fiction. Names, characters, places, and incidents either are the product of the author's imagination or are used fictitiously, and any resemblance to actual persons, living or dead, business establishments, events, or locales is entirely coincidental.

Library of Congress Card Number: 2007902316

ISBN 978-0-9795001-0-7

Printed in the United States by Morris Publishing
3212 East Highway 30
Kearney, NE 68847
1-800-650-7888

Acknowledgements

Praise God from whom all MY blessings flow! It's been along time coming, but I finally got here…but not all on my own! First and foremost I have to give thanks to God for bestowing me with the love of the written word and the courage to use it to inspire others. A couple of special shout outs to those who helped in making this book possible. Thanks to my mother for simply being my mommy! Ike, my BFF, thank you for seeing the vision and believing in me especially on the days when I didn't believe in myself. Wanda, my oldest and dearest friend, for cheering me on (that's my girl on the front cover). To all of my sista friends…Kym, Trell, Sophia and Audria for reading and rereading chapter after chapter without growing weary. A special thanks to Lisa N. for keeping me on the straight and narrow and who unselfishly helped to bring my dream to fruition. Tracy for all of the pep talks that kept me uplifted. My cousin Tina for getting the word out! The world's best editor, Judy Allen of Marketing Solutions. Thank you to Stephanie Jackson of Total Paradise for keeping me looking fly and getting on the bandwagon at the first mention of PYWT. K.D. for rescuing me! The pictures are slammin'! To Eva for believing in PYWT before the first word was ever written. I thank God for you and all of your advice, support, encouragement, prayers and pearls of wisdom. Lastly, to my two favorite guys, Darrion and Isaiah. Mama loves you. Everything that I do—I do for you!!!!

In loving memory of my dad
James Franklin Cornelius
1948 – 2002

Chapter One
Tamara

I've learned not to question some things in life, because life is a mystery and in time the truth shall reveal itself. Take my life, for instance. I'm a 35-year-old, successful, intelligent, beautiful, single woman with no children. Now, why I'm still unattached is a mystery to me! Because if I say so myself, a sista got it going on! Now, I've had a number of guys ask me to marry them. But it's always been from the wrong guys. You know the type. They're either in between jobs, car-in-the-shop, living-with-my-mama-to-help-her-out, no-future type of guys. All losers. But then I met Samuel.

We met in a coffee shop around the corner from where I lived. I stopped there every morning before going to work. In all the years that I've been coming to Sugar & Spice coffee shop, I've never once noticed him. For the life of me I don't know how I have missed him. He was 6 feet, 4 inches of deep dark chocolate! From the moment our paths collided (quite literally, by the way) I've been head over heels.

I was rushing into the coffee shop, late for work as usual, when I ran right into Samuel and his latte. After spilling his coffee on the front of my favorite white crocheted sweater, he apologized profusely and insisted on paying to get it dry-cleaned. So we exchanged business cards. I walked out of the coffee shop totally in awe. From the outside he was absolutely gorgeous and from the brief conversation we shared, he appeared to be the same on the inside as well. Could this have been fate?

Just think. This morning I almost skipped my routine coffee shop visit because of deadlines at work. But I couldn't resist my usual hazelnut latte with a sprinkle of cinnamon, and frankly I just love the way the place smells. What if I'd have ignored that voice inside my head screaming *"Coffee"* and went straight in to work? I would have never met Samuel. Life Lesson #1...Our steps are ordered!

With a coffee stain the size of Mount Fuji covering the front of my sweater I truly looked a mess and needed to change before going into work, which was going to really make me late.

Using my cell phone, I called into the office to let my assistant know that I was running behind and to push back my meetings. I hated to have to do that especially since my first meeting was with Clyde Peterson. He was the Director of Human Resources and was known to be difficult at times. But he also had a weakness for the ladies, so I'll just have to make it a point to be extra sweet to him. How I detest the thought! Though some women regarded Clyde as very handsome, he definitely was not one of my favorite people.

Stepping off the elevator, I was in a hurry to get his day finally started and I brushed past Sheila, my assistant.

"Well, good morning to you too, Ms. Thang," said Sheila.

Sheila Wilcox had been my executive assistant for three years. She was assigned to me when I started my position as Director of Benefits with AllState Insurance Company. From day one we instantly hit it off and have been close ever since. I don't know what it was, but there was just a presence about Sheila. Maybe it was the huge Kool-Aid smile that always adorned her face, or her sunny disposition. She had a way of making you feel like you've known her all your life. I had grown to love her as the sister I never had. We looked nothing alike. I was much taller, even with the three inch heels that she wore everyday. Sheila was considered pecan tan and was without a doubt as cute as a button.

"Sorry Sheila," I replied. "I've already had a heck of a morning and it's not even noontime. Did you get everything squared away for me?"

"Don't I always?" Sheila answered.

"Yes, you do. Remind me to send you some flowers," I said as I walked toward my office.

"I prefer cash," Sheila responded.

"How about lunch?" I asked.

"That'll work too. You know I'm easy," she announced with a laugh.

"Well, pencil yourself in for today."

"Will do," Sheila commented.

As I approached my office, I opened the door and saw the largest floral arrangement I'd ever seen covering my entire desk! Now I had male fans, but no one that I was seeing seriously enough to send flowers.

"So is that why you were running so late this morning?" Sheila asked, now standing directly behind me.

"Who are they from? I know you've already read the card," I asked excitedly!

"Well, you know I had to look out for my girl and make sure everything was on the up and up," Sheila justified.

"Ummhmmm…just being nosey is more like it," I joked half-seriously.

As I opened the card, my mind was racing, wondering who in the world would be sending me such a beautiful arrangement. The card was signed "Samuel Roberts". My breath was taken away. Mr. Latte had sent me flowers.

"So who is he? I've never heard you mention him before," Sheila asked.

"Oh my goodness. I'm not believing this." I was astounded. I have never believed in love at first sight, but this man had definitely gotten my attention.

"Well, believe it! Now spill the beans! Who is this Samuel Roberts?" Sheila asked, growing impatient.

"I just met him this morning, and hear me when I say brotha was fine! I bumped into him at the coffee shop -- literally. He spilled his coffee all over the front of my white sweater."

"You mean the crocheted one that I've been dying for?" Sheila asked.

"Yes, that would be the one," I told her.

Sheila gasped and shook her head in disgust. "Keep going."

"Well, after spilling his coffee all over me, he was adamant about getting it dry-cleaned. So we swapped numbers. I, then, had to rush home to change clothes and that's why I was late this morning, Ms. Nosey!"

"Since when does spilling liquid on the front of a woman's blouse warrants flowers," Sheila asked sarcastically. "Very impressive. I like him already," she said with a devilish smile.

"I think I do, too." I couldn't help but smile to myself at the thought of Samuel's dimpled grin. We both jumped as someone knocked at the door. It was Clyde.

"My…my… someone has been a good girl," Clyde replied.

"And by the size of that bouquet I'll say a very good girl indeed."

"Well, I have work to do. Nice to see you Clyde," Sheila said as she walked out of the office. Clyde turned to watch as she exited the room.

"It was very nice to see you also, my dear." Clyde Peterson was always the charmer. But not even he could lower me from my cloud this morning. My mind was filled with thoughts of Samuel, my potential new boyfriend.

Clyde was wearing what appeared to be a very expensive suit tailored to fit him perfectly. I had to admit, only in my mind, that he was looking quite dapper. Clyde had a cinnamon tone with chiseled features and a deadly stare. You could tell even through his suit jacket that he was in good shape. While he was not very tall it didn't seem to keep him from getting what or who he wanted.

"Clyde, thank you so much for allowing me to push our meeting back, I really appreciate it."

"No problem. You had a late start this morning." He asked as he looked over at the bouquet.

"Yes, I guess you can say that. But I'm here and I'm ready for business now," I commented as I pushed the floral arrangement to the side to make space on my desk.

"Good. Did you get a chance to look at the new insurance packet?" He asked now in a very business-like tone.

"Yes and everything looks good."

"You can say that again, because from where I'm sitting everything looks exceptionally well," he said flirtatiously. So much for business. Clyde's playfulness with words went on throughout the duration of our meeting. After about twenty minutes, I had had my fill of him and was ready to bring the meeting to an end before he ruined my high.

"Well, if that's all Clyde, I'll have Sheila enter the numbers and the changes will be reflected on the next pay period," I commented as I stood, and extended my hand to him signaling the close of our meeting. "Again, thank you for your time."

Clyde slowly rose from the chair he was sitting in, took my hand and kissed it. "The pleasure was all mine, Ms. Washington." Clyde walked out and winked as he closed the door behind him.

Men! You can't live with them and you sure can't shoot

them! I sat wondering if his wife knew how blatantly flirtatious he was. What was I thinking? Of course, she does. She married him and I'm sure that's the way they met. But silly her probably thinks she has reformed him. But like the old saying goes, 'A leopard doesn't change his spots'. Once a playa always a playa. They don't die, they just get old. Ain't nothing sadder than an old playa who doesn't know that he has played out. I don't trust any of them as far as I can throw 'em. That's why I will never let one tie me down. Just love 'em and then leave 'em, before they leave me. That's my motto.

But this Samuel intrigued me. Could he be different than all the rest? There's only one way to find out! I had just enough time before my next meeting to give him a call to thank him for the beautiful, fragrant gift. I grabbed my purse and started to rummage through it. What did I do with his card? I really needed to clean this thing out. Here it is. *Roberts' Accounting Service*s. *He's an accountant and owns his own business. Good looks and brains. What more could a girl ask for?* I'll just give him a quick call.

I picked up the phone and dialed his number. It was ringing. I could actually hear my heart pounding. *Why am I so nervous? He's just a man.*

"Roberts' Accounting Service. May I help you please?" The receptionist on the other line asked.

"Yes, may I speak with Samuel Roberts? You can tell him that Tamara Washington is calling," I said as smoothly as my fiercely pounding heart would permit.

"Hold please, Ms. Washington," the receptionist responded very crisply.

"Thank you."

A few seconds later I heard his sexy baritone voice.

"Hello, this is Samuel."

"Good Morning again. This is Tamara Washington from the coffee shop."

"Yes, Tamara. How are you?"

"I'm doing fine, great in fact. I just wanted to call and thank you so much for the lovely flowers, I absolutely love them. But you didn't have to do that."

"Of course, I didn't have to. But I wanted to. I felt so bad for spilling my coffee on your sweater. Are you sure you're okay?" He sounded genuinely concerned.

"Yes, I'm fine," I reassured him.

"Be sure that you give me the dry cleaning bill for it and I'll make sure that it's taken care of. That's the least I can do," he insisted.

"Okay, if you insist. But it's really not necessary."

"Then it's settled," he said.

There was a knock at the door.

"Well, Samuel I was in between meetings and had a minute to call to thank you. So thanks again. It really brightened my day."

"I'm glad that I could do that for you and maybe I'll get the opportunity again. To brighten your day that is," Samuel said.

"Yes, maybe you will," I smiled.

"Well, I'll have to see what I can do to make it happen," he said eagerly.

"You do that."

"Thanks for calling, Tamara. Now you have really brightened my day," he flirted.

"Well, I'm glad that I could do that for you and maybe I'll get the chance again some time," I flirted back. Just then Sheila buzzed me on the intercom.

"I look forward to it. Just don't let it be too long," he said practically begging.

"Okay, but I have to run. Thanks again."

"You are most welcome."

"Bye Samuel."

"Until we meet again, Tamara."

I hung up and I was in a daze. *Was it my imagination or was he doing some heavy duty flirting with me? Lawd have mercy!* What a sexy voice! It had me captivated.

Sheila buzzed again and brought me back to reality.

"Umm yeah."

"Tamara, is everything okay in there?"

"Yes, I'm sorry. I was on an important business call," I lied.

"You're next appointment is waiting. Are you ready?"

"Yes, Sheila. Send them in please." It was time to get back

to work.

I had just finished my third consecutive meeting and was starving. I walked out of my office to find Sheila talking to Clyde. He was sitting on the edge of her desk and she was listening attentively to whatever it was he was saying. He stopped talking and stood up to leave when he noticed me walking toward them. Now my curiosity was piqued. What in the world was he prowling around for?

"See you on Sunday, Sheila," Clyde said.

"Okay and congratulations again," Sheila beamed.

"Thank you. You should bring Tamara with you on Sunday," he commented over his shoulder as he entered the elevator. He smiled slyly at me as the elevator doors closed.

"Bring me with you where?" I asked while standing over Sheila.

"To church," Sheila answered. "He has just accepted the position as the new senior pastor at my church."

"Pastor! Clyde! You're kidding right?" I bent forward laughing.

"No. Why do you say that?" Sheila asked, sounding offended. I stood back and took a good look at Sheila. She was dead serious.

"Because I can't believe the flirtatious, womanizing, egotistical Clyde that I know is a pastor of a church?"

"Tamara, being a pastor doesn't mean he's perfect. He's still a man," Sheila commented.

"Still a man! I don't know about you, but where I come from being a pastor is supposed to mean something." Realizing I was raising my voice at Sheila, I regained my composure. "I'm sorry, Sheila girl. I didn't mean to get loud on you. I guess those meetings stressed me out a little more than I thought. I don't mean to take it out on you. You ready for lunch? Because I'm starving."

"Yeah, whatever. I made us reservations at Taylor Made Grill Restaurant for twelve, but now we're late," Sheila said obviously irritated.

"Well, let's get rolling. I'm so hungry I could eat everything on the menu right about now," I said joking trying to clear the now stale air. "You driving, right?" I asked Sheila.

"No, you're driving. I drove us the last time. Remember?"

"Are you sure?" I asked innocently.

"I'm positive. So get your keys girlfriend," Sheila said now smiling.

Taylor Made Grill was crowded, and because we were late we had to wait 20 minutes for a table. The aroma of their freshly baked breads engulfed me. My stomach had stopped growling and was now screaming for food.

"Could you order me my usual? I have to make a quick dash to the little girl's room?" Sheila asked.

"Sure."

By the time Sheila had returned, our salads had been served.

"Mmmm…I just love their bread," Sheila said stuffing her mouth full of Taylor Made Grill's famous Asiago Cheese bread.

I just sat there staring waiting for her to bring up our prior conversation regarding the weirdness of Clyde being pastor at her church. She finally looked up and noticed me staring at her.

"What?" Sheila asked. "I thought you were so hungry. Is something wrong with your salad?"

"No, there is nothing wrong with my salad. Now you spill the beans! Since when is Clyde a pastor, and at your church no less?" I asked.

"Good Lord, Tamara. You're still on that? You know if you took the time to get to know him, you'll find that he's a really nice guy. I've talked with him on numerous occasions regarding issues with Leon and me. He's given me some really good advice," Sheila admitted.

I couldn't believe my ears. "You mean to tell me that you let another man into your private, sacred marital affairs?" I exclaimed.

"He's not just another man, Tamara. You have to stop looking at him like that. He's a minister," she clarified.

"Yeah right, he's a minister. You just watch yourself. The devil comes in all disguises," I shot back.

"Tamara, you don't know him like I do. He's really sincere."

"Well, let me ask you this. Has he talked to Leon or is he just talking with you?" I asked.

Sheila didn't have to open her mouth to answer my question. The look on her face told it all.

Chapter Two
Sheila

As I drove home from work, I wondered, when did things start to go wrong between Leon and me? We were once so much in love. There was a time when you seen one of us, you seen the other. We were just as the song said, *Inseparable.* I just couldn't seem to get enough of him. Now lately, a few minutes in the same room was too much to bear. Just the sound of his breathing got under my skin and just like the other song said, there truly was a thin line between love and hate.

When I first met Leon, I wasn't at all attracted to him. He was tall enough and cute enough, but he also was a bit rough around the edges. You could tell he had been around the block a time or two. He was a little on the skinny side and kept his head shaved. A friend of mine introduced us. He was coming out of a relationship and I was eager to begin one. I can confess now that I was a bit intimidated by him. He was in a word…"seasoned". What I mean is that he was no stranger to having relations. I, on the other hand, was very inexperienced, but I threw caution to the wind and went out with him. One date turned into a whirlwind of others. Leon proved to be harmless. He was just a man looking for love, which I openly gave. I soon thought I had found my knight in shining armor. He made me feel safe...protected...loved. Nothing bad could happen to me as long as I was with him. He was my security; he's always there when I needed him. Unlike my father, who was never around because he was too busy chasing women. No, Leon was different, so I thought. So what happened to that man I used to adore? Or maybe it wasn't him that I really adored, but an image of what I wanted him to be. Either way the thrill was gone.

Today was Leon's day off, so I didn't have to rush home. On such occasions I usually worked late or ran errands. He would pick the kids up from school and hopefully would feed them too. Since going home was the last thing on my mind, I thought I would pay Mama Lyn a visit. Visiting with my Mama Lyn always made my spirit feel good and my spirit was uneasy ever since lunch. Tamara's question still lingered on my mind. *Why hadn't Clyde suggested talking with Leon?*

I was so relieved to see Mama Lyn was at home and not off somewhere visiting the sick or up in some store shopping. Mama Lyn knows she loves herself a good sale. She had more shoes than DSW. Though she had gotten up in age she still loved to get her shop on. As I entered the house I found her in one of her favorite places. I simply followed the delicious aroma.

"How's my favorite grandmother?" I said sneaking up behind her.

She turned around smiling, not even startled to see me. "You mean your only grandmother," she embraced me tightly. "You've been heavily on my mind, chile. How is everything?" Mama Lyn was short and hippy. Her tan complected face constantly displayed a smile and for as long as I could remember she always wore her silvery, long hair in a fashionable bun.

"Oh, everything's fine," I said avoiding eye contact.

"Where are the children?"

"They're with Leon. It's his day off."

"That's right and how is grandson-in-law?"

"He's fine, Mama Lyn."

"Can I fix you something to eat? I was just touching up my greens. I could use a second opinion."

"By the way they smell, you have already touched them up just right Mama Lyn."

"Let me get my cornbread out the oven and I'll fix you a plate."

That's another good thing about visiting Mama Lyn. She was always going to feed you. Both your spirit and your belly would be full after leaving her. She placed a big helping of greens and hot buttered cornbread in front of me. As I began to eat, she poured me some ice tea. I took a bite of my food, which to no surprise was lip smacking.

"I can't believe I caught you at home."

"Actually, I was up early this morning. Macy's had a one-day sale from 7-11 am and I wanted to beat the crowd. I was there before the doors opened, with my charge card in hand," she chuckled.

"You heard that Clyde, I meant Reverend Peterson,

accepted the position of pastor at the church?" I asked Mama Lyn.

"Uh huh," Mama Lyn said as she took a seat at the table with me.

"I'm excited. He seems to be a good man."

"Uh huh," Mama Lyn commented while looking at me suspiciously.

"I met his wife and she's very sweet."

"Now chile, you know that Mama Lyn knows you, and I know you didn't come all the way out here to talk about Reverend Peterson and his sweet wife. So tell me what's really troublin you."

I exhaled and sat back in my chair.

"Mama Lyn, do you think that Reverend Peterson is a good man?"

"Well, I can't say that I actually know him that well. I haven't talked to him much."

"But what do you feel?" I asked her.

Now Mama Lyn sat back in her chair. "I feel he has the potential of being a good man."

"What does that mean?"

"What I mean is he has been called to be a good man, but as to whether he is open to letting God totally lead him is up to him. You see honey, many are called, but few are chosen. Being a pastor is a tough job, it's nothing to take lightly. The man of God has to stay prayed up, because the devil is always on attack. He is going to throw any and everything in his way to try to make him fall. He will use whatever tactic to distract him from doing God's work. So, we have to stand in the gap and pray for them as well. Especially for Reverend Peterson, because that man has a lust demon riding his back!"

I was shocked at Mama Lyn's last statement. Did she know something that I didn't know about Clyde?

After leaving Mama Lyn's, I was even more confused about Clyde--Reverend Peterson. *Was he really a womanizer? Did he really care about my martial problems with Leon or did he have ulterior motives? And why does my heart skip a beat whenever he's near?*

I loved our home. We lived in Rivoli Downs, which was a gated community located on the north side of Macon. We bought the house five years ago when I was pregnant with my youngest

Indya. It was a European-style two story house with vaulted ceilings and hardwood floors throughout. It was decorated in rich earth tone colors to evoke a sense of relaxation. But as I walked through the door everything was chaotic. Indya was still up running around, her toys were scattered everywhere, and I didn't detect the aroma of food coming from the kitchen. Nope, just the same dirty dishes that were in the sink this morning. What a welcome home! This was enough to make me go postal.

"Mommy's home! Mommy's home!" shouted Indya. She ran up to me, jumped into my arms and gave me the biggest hug. It felt so wonderful. Indya was my very own, living, breathing baby doll. She had big eyes that gave her oval, honey colored face an innocent look which worked to her advantage many times when discipline was called for.

"How's mommy's sweet baby?"

"Good! I got a smiley face on my papers today. Wanna see 'em mommy?"

"Of course, sweetie. Where is everybody?"

"Kelsey is in his room playing on that game and daddy's in the backyard." Kelsey was my nine-year old son and was the spitting image of his father—minus the goatee, of course.

"Have you eaten?"

"Yep. I had a peanut butter sandwich and I fixed it all by myself," Indya said proud of her accomplishment.

Just then Leon walked in from outside, surprised to see me. "Hey honey. I didn't know you had made it home yet," he pecked me on the cheek. "I was just about to get the kids ready for bed."

I looked at my watch. "They should have been in the bed by now, especially Indya," I snapped at him.

"I know, but I was tinkering with that old car of mine and time got away from me."

"Ummhumm."

"You had to work late?"

"No, I went by to see Mama Lyn."

"Oh, is your grandmother doing okay? She's not sick is she?"

"Yeah, she's fine. Doing better than me," I mumbled under my breath.

"Whachoo said, baby?"

"Nothing. Nothing at all, Leon."

"Did you already eat? Would you like for me to fix you a sandwich?"

A sandwich! What kind of dinner was that to come home to I thought to myself. If I would have been home all day he would have had a hot-cooked meal waiting on him. But I guess that's too much to ask of him to do for me. A little too much like right!

"No thanks, I had something at Mama Lyn's."

He walked toward me and tried to give me a hug. I moved away.

"Come on, Indya. I guess mommy will get you ready for bed," I smirked.

I left him standing there alone, exactly the same way I felt. *Was I in this marriage all by myself?* After I got both the kids tucked in, I was ready to hit the bed. It had been a long day and I was exhausted. When I went into the bedroom, Leon was just getting out of the shower. He walked towards me and I turned away.

"What's up with you?" he asked.

"What do you think is up with me?"

"Hell, I don't know! You came home with a chip already on your shoulder!" he said angrily.

"I'm too tired and don't feel like getting into it with you tonight, Leon."

"Well, that's just too bad because I do! Sheila, I work hard all day and place my whole paycheck in your hands. You can have anything you want, but you still seem to not be happy. What more am I supposed to do?"

"You think money is the answer to everything! I don't care about money! For once I would like to come home and have everything laid out for me! You get home from work and don't have to do nothing! I cook for you, keep the house clean and take care of the kids. That's outside of my eight-hour a day job! When was the last time you cooked dinner for me? When was the last time I was made to feel special?"

He just stood there in silence with nothing to say, dumbfounded. Just as I thought. I went into the bathroom to run a bath. I needed to soak and pray. As I sat in our Jacuzzi tub, I relaxed as the jets bubbled around me and I cried out to the Lord.

"Lord, I know you made me to be my husband's helpmate and I do that wholeheartedly. It brings me great joy to take care of the needs of my husband. I feel that's my duty as his wife, but what about my needs? Is it right that my happiness is forfeited? Aren't I important too? Don't you care that I'm hurting and in need of comforting. It's been so long since I've felt loved and appreciated that I fear I'm slowly slipping away. I'm afraid of these feelings and of what I may do. You said that if I delight myself in You that you would give me the desires of my heart. My desire is a man who loves me as Christ loved the church. Father, please come see about me...before it's too late."

Leon walked into the bathroom to brush his teeth. "Are you coming to bed soon?" he asked.

"Yes, I'm just going to soak a little longer. Could you close the door behind you, please?" After he left, I felt compelled to whisper yet another little prayer.

"And Lord please let Leon be fast asleep, by the time I get out of this tub."

I guess the Lord was busy with other prayer requests, because he hadn't heard my cry. Leon was still up and waiting. After I lotioned my body, I put on a pair of flannel pajamas. But old lover boy still didn't get the hint. As soon as I was under the sheets, he was all over me like white on rice. That peck on the cheek earlier when I first arrived home was all the foreplay I was going to get for the night.

"Lord, it's me again. Please help me to get through this!"

Chapter Three
Tamara

I woke up in an excellent mood. It was one of those mornings where you can feel something good was going to happen. It was a beautiful spring day and I was enjoying the soft breeze as I was driving with my convertible top down. The temperature was perfect and the cherry blossoms were in full bloom. As I drove towards Sugar & Spice for my morning coffee, I hoped that Samuel would be there. It would be an added bonus to my already good day.

I found a parking spot right in front of the coffee shop. I stepped out of my car wearing my caramel colored tunic blouse with the matching slit skirt and my caramel colored high-heeled sandals perfectly displaying my French manicured toes. I looked good. As I walked into the coffee shop, I nonchalantly scanned the room looking for Samuel. I didn't see him. Darn! My heart dropped a little, so I patiently waited in line with the rest of the coffee lovers. I was next in line when a sexy Barry White-sounding voice startled me from behind.

"Good morning, beautiful."

I turned around and the handsomest man since Denzel was standing before me—Samuel. I struggled to maintain my self-control, but I couldn't suppress a smile. He was so scrumptious looking.

"Good morning."

"I was hoping to run into you this morning—minus the coffee," Samuel said. We both chuckled.

"Really now. Well, it's a good thing I'm in front of you in line or who knows what would happen," I joked.

"Oh believe me, I have learned to be more careful while walking with a beverage in my hand."

"Well, that's good to know."

"May I take your order, ma'am?" The cashier asked with a slight irritation in her voice. *I forgot where I was. How does he do that to me?*

"Yes, I'm sorry. I'll have a cup of a hazelnut latte with a dash of cinnamon to go, please."

"Add a French vanilla latte too, please?" Samuel told the cashier. "Would you like anything else?" Samuel asked me. He was standing so close that I could smell his mouthwash. I hoped the mouthwash that I had used this morning was what he smelled from me, too.

"No, thank you," I blushed.

"Let me get a dozen of your almond flavored biscotti."

"Will that be all?" the cashier asked Samuel. He looked over at me once more before answering her. I nodded.

"That will be all." He opened up his wallet, pulled out a platinum card and paid for our order. Baby had bank! The cashier handed us our lattes and Samuel the bag of biscotti. Then he handed the bag to me.

"For you and your co-workers."

"Well, thank you," I said still smiling.

"No, thank you."

"Why are you thanking me?"

"For being here. That's the only reason why I came by this morning. I can get coffee anywhere. There's a Dunkin Donuts on every corner."

I was grinning from ear to ear.

"Tamara, would you please allow me to take you to lunch today?"

I wanted to leap in the air and click my heels together, but I didn't want him to think that he had it like that.

"I'm flattered, but I will have to check my schedule. Can I give you a call later and let you know if I'll be able to go?"

"Of course, please do."

If I didn't know any better, I would have thought he was a bit disappointed. I guess he's used to women falling at his feet. *Not me, buddy. It's going to take a mighty big bait to catch this fish.* He walked me to my car.

"Your keys."

"Huh?"

"I was asking for your keys, so that I could open the door for you."

"Oh yes, my keys. Here you go." I felt like a total doofus.

He opened the door and I slid into the front seat.

"Thank you."

"My pleasure. I'll be waiting for your call."

I smiled and drove off. Yes, this was going to be a good day! I believe I floated into the office, because I didn't even remember driving in.

"Good morning, Sheila!" I beamed.

"Morning," Sheila responded dryly.

"Isn't it beautiful out today?"

"Uh huh."

I overlooked Sheila's very obvious bad mood for now. I had to hurry into my office to check my schedule for today. I turned the computer on and accessed my calendar. Lunch was wide open. *Hallelujah! I'll just wait a few minutes before I call Samuel. I don't want to seem too anxious. Even though I am!* For now I'll get back to Sheila and what was bugging her.

"Hey you," I said to Sheila.

"Hey," Sheila answered not even looking up from what she was doing.

"Are you okay?" I asked concerned.

"No, but I will be."

"Anything I can do to help?"

"No, but thanks," Sheila said giving me a half smile.

"Don't give me that. You know you can talk to me Sheila. We're not just co-workers, we're friends...sister friends. So spill the beans. What has Leon done now?"

"Nothing, Tamara. It's nothing," Sheila said rolling her eyes.

No, she didn't just give me attitude! Well I tried, but ole girl doesn't want to open up. If she didn't want to talk about it, I wasn't going to twist her arm. So I jumped right into my early morning event.

"Okay. Well, guess what happened to me this morning?" I asked Sheila turning a blind eye about what was obviously going on with her.

"I give up," Sheila said not enthusiastic at all.

"Samuel was at the coffee shop and he asked me to go to lunch with him today!" I exclaimed. "Oh, he bought biscotti, too. I had forgotten. It's still in my car. Let me run down and get it.

I'll be right back to fill you in on the rest."

I practically skipped to the elevator. As I was entering the elevator, Clyde was stepping off. *Now what is he doing on this floor and who is he coming to see? Is he was coming to see Sheila? Maybe she called him to talk to instead of me. I guess she feels more comfortable talking to her new pastor than me these days.* I can't quite put my finger on it, but something just doesn't feel right about this newfound relationship between Sheila and Clyde. I know she says she only confides in him about her and Leon's martial problems, but what's in it for him? What is he really after?

I don't care much for Clyde. Not very much at all! He had a sound business mind, but it was also just as perverted. There had been plenty of rumors circulating about him and a few of the female employees. He could always be seen in some woman's face, whispering and smiling, off in some corner. He definitely had a weakness for their sweetness, and they seemed to have a weakness for him as well. He hadn't approached me directly, but I had experienced his loose tongue during a number of one on one meetings together. I ignored his crude comments, even though several of the women appeared to enjoy his advances. Clyde had what some folks would call "charisma" and he knew how to work it to his advantage. He was as sly as a snake and just as deadly, but I wasn't affected by Clyde's cunning ways. I had dealt with men like him before, sadly in my very own family.

I had an uncle who just couldn't seem to keep his hands to himself or off of me. Uncle Ray was a slick one too. He befriended me, found out my vulnerable areas and then used them against me for his own selfish reasons. I was tall, lanky and dark-skinned; to most people not too attractive, except to my Uncle Ray. He said I was a black Nubian queen with natural beauty that I should be proud of. He would come around when I was alone and we would sit and talk for hours. He reassured me of how pretty I was, warned me about the boys at school, bought me nice things and even slipped me a little spending money. I was his special niece and truly started to feel like a Nubian queen.

Then one day during one of his visits he gently rubbed against my breasts. I assumed it was done by accident and didn't object. I guess he took my silence as permission to go further. So the next week he kissed me. Open mouth! I was startled to say the least, and pushed him back. That only pissed him off. He called me a tease and grabbed me by my shoulders. He shoved me down on the sofa and forced himself on me. I tried to fight him off, but he was too strong. I was scared. Petrified. I kicked and screamed, but that only made him madder. He covered my mouth with his big paw and pressed down even harder. I struggled underneath him, but was no match. I surrendered, powerless to his strength. He glared down at me as I cried. He watched as tears softly ran down the sides of my face. When he was finished, he whispered in my ear that he loved me. I had an overwhelming urge to throw up. Before he left he told me I had better not tell anybody and that no one would believe me noway.

Long after he had gone I still laid there motionless, until my mom arrived home. Despite his warning I told my mother what had happened. I told her everything. But just as he said, she didn't believe me. No one believed that the good, right Reverend Ray could do such a thing. From that day forth I never set foot in a church again. Come to think of it, I don't suppose I prayed much since then either. I called out to the Lord that awful day, but He didn't answer. How could a God so loving allow that to happen to a child? I was only thirteen years old!

Chapter Four
Sheila

"Hello, sweetness."

"Clyde. I mean, Reverend Peterson. I didn't hear you walk up."

"Sheila, we are at work. I'm not the pastor here. Please continue to call me Clyde. "

"Okay, Clyde." My heart was racing. He was looking and smelling good enough to eat. He sat on the edge of my desk and was so close to my face I thought he was going to kiss me. Secretly, I wished he would. *What was coming over me?* He placed his hand under my chin.

"Why the long face?" he asked.

"Oh, it's nothing," I looked away.

"Sheila, I came all the way over here just to see your lovely smile and it's not here. So tell me what's wrong. What or who has stolen my smile?"

"I hate to keep bothering you with my problems."

"Nonsense. That's what friends are for and we are friends, right?"

I nodded my head.

"So talk to me."

"Okay, but not here."

"That's fine. Why don't you come to my office? We can have all the privacy we need there, with no interruptions."

"Okay."

"So, I'll see you in about half an hour and don't forget to bring that beautiful smile with you."

"We'll be there," I said smiling for the first time that day.

Clyde bent down and kissed me on the cheek. An alarm inside of me went off! He then smoothly walked off toward the elevators. He stood confidently inside the elevator smiling at me until the doors closed. A part of me was turned on by it all, while another part of me wanted to run and hide for cover.

I was not sure what it was about Clyde, but he awakened a deep part of me that has been lying dormant for years. He made me feel pretty and sexy—like a woman. Being married to Leon for so long, I had forgotten what that felt like. Leon sees me as just the mother of his children and someone to provide occasional sexual relief. I received no romance, no gifts, no flowers, no nothing. My life with him was slowly driving me insane. It was the same boring, grinding routine day in and day out. Clyde was the only pleasure I had to look forward to. I realize that it is wrong to have feelings for any other man than my husband. But what is an emotionally staving woman supposed to do? Tamara had returned from her car with a bag of what I assumed was the biscotti in her hand.

"What was Clyde doing here?" Tamara asked.

"He just stopped by."

"Uh huh," She looked at me skeptically.

"Excuse me, Tamara. But I need to go to the little girl's room."

"Okay. Sheila, are you sure you are alright?"

"Yes, don't worry. Everything is going to be just fine," I smiled at her.

I dare not tell her that I was going to see Clyde not after the way she reacted the other day. No, I'll just keep this little secret to myself. It's not good for a girl to tell everything she knows. Never let your right hand know what your left hand is doing. That's what my Mama Lyn always says.

Now Tamara was my roll dog and all, but she just wouldn't understand. She wouldn't understand giving all of yourself to a man and getting nothing back in return. She wouldn't understand wanting the man you love with all your heart for once to go out of his way for you...just because. No explanation, except the fact that he loved you. Isn't that what love is all about anyway. Doing what would make the other person happy, even if you don't feel like it. There are plenty of days when I didn't feel like coming home from a hard day on the job and cooking dinner. But I did it as one of my ways of letting Leon know I appreciated him. Then there are nights when I'm not in the mood for having sex, but because of the love I have for him I obliged. Some nights I just want to be held until I fall asleep. Cuddled.

I would do anything to feel valued as a woman and not just a piece of meat. Leon has gotten way too comfortable in our marriage, to the point that he doesn't deem it necessary to wine and dine me anymore. In the beginning I was his queen, now I'm feeling more like the hired help. But that's not the way Clyde made me feel. He moves me in a way Leon used to, before he had become so complacent. I know it's wrong to have these desires for Clyde, but it feels so good and it just feels good to be able to feel anything again.

As I was walking away, I heard my phone ringing. Whoever it was will just have to wait. I was a woman on a mission. I did make a stop by the ladies room to freshen up a bit. I wanted to make sure everything was in order and not a hair was out of place. I even sprayed a little perfume on and checked to see if the dress I was wearing showed just enough thigh. I have been told that my shapely legs were one of my most valuable assets. I was going to let them work for me today.

Because of my nervousness the walk to Clyde's office took an awfully long time today. I guess he told his secretary he was expecting me, because she motioned for me to go right in. Surprisingly, when I entered inside his office he had it dimly lit. *Was conversation the only thing on Clyde's mind? Was he feeling the same thing that I was feeling?* Only time would tell.

Chapter Five
Tamara

I don't understand why Sheila keeps brushing me off. We usually talk about everything—from my crummy love life to PMS. So, I wondered what was going on with her, and why Clyde was hanging around so much. I'm not sure if anything is going on between them, but something just doesn't feel right. Then again, maybe it has something to do with Leon and she's just confiding in Clyde. Whatever is going on I hope she is using her head, because Clyde seemed to be trouble with a capital "T" to me. I just don't trust the man, even if he is calling himself a minister these days.

I believe I kept Samuel waiting long enough. I'll call him to confirm our lunch plans for today. Even though I don't have much of an appetite, it will be nice just to sit in his presence. I don't think I have felt this giddy since high school and that's been a very long time!

"Roberts' Accounting Service. How may I help you?"

"Samuel Roberts please."

"Who may I tell him is calling?"

"Tamara Washington."

"Hold please, Ms. Washington."

A moment or two passed.

"Ms. Washington, I'm sorry but he is on another line. Would you like to continue to hold?"

"No, just tell him that my schedule is clear and lunch today will be fine. I will make reservations at Taylor Made Grill for 12 noon."

"I'll be sure that he gets the message."

"Thank you." I hung up the phone and danced a jig. I walked out of my office to tell Sheila my news, but she still was missing in action. Her phone was ringing, so I answered it. "Benefits Department. May I help you please?"

"Sheila?"

"No, Sheila has stepped away from her desk. Is there something that I could help you with?"

"Tamara, is this you chile?"

It was Mama Lyn!

"Mama Lyn!" I exclaimed. "I'm so sorry. I didn't even recognize your voice. How are you?"

"Mama Lyn's doing just fine. I can't complain. God is good. How are you doing baby?"

"I can't complain myself and I guess it wouldn't do any good, because nobody cares anyway," I joked.

"Well, that's not true. Jesus cares," Mama Lyn commented.

I could feel a sermon coming, so I tried to rush past her. Now, I adored Mama Lyn and loved her just like she was my own, but I just didn't want to hear how good God was today. I was excited and wanted to tell anybody who would listen about Samuel!

"Did you want me to give Sheila a message for you?"

"Just tell her I called. After our last visit, she's been on my mind. Something seemed to be troublin her and I wanted to let her know that Jesus is able and that I'm praying for her."

"I sure will, Mama Lyn."

"Thank you baby and Mama Lyn's praying for you, too."

Praying for me! Now that knocked me for a loop. What in the world was she praying about for me?

"Thank you, Mama Lyn," I muttered.

"I love you," Mama Lyn expressed to me.

"Love you, too," I said in a whisper almost too low to hear.

So it just wasn't my imagination, Sheila was acting strangely. Mama Lyn noticed it too. Just as I was putting the phone back on its cradle and turned around, Leon was standing right behind me.

"Hey there, Tamara. How's it been goin', girl?" Leon asked smiling.

"Hey Leon. It's goin'," I commented, startled to see him.

"So, where's my beautiful wife? I wanted to surprise her and take her out to lunch."

"How thoughtful of you, but she isn't in right now. She's in a meeting," I lied.

"Oh, really. She never mentioned a meeting to me, but then again, she hasn't been mentioning much lately," he admitted sadly.

"I'm sure it just slipped her mind," I reassured him. I had to get him out of here fast. I didn't know when Sheila would arrive back to the office from her mysterious whereabouts. "I was on my way out myself, but I'll leave her a note that you came by."

"Oh, I didn't mean to hold you," he said.

"You're not. It was good to see you again. You should stop by more often."

"Yeah, right. For all the good it did," he commented with sadness in his eyes. He left as quickly as he came.

I suddenly felt sorry for him. I liked Leon. He had his faults as we all do. But he was truly a nice guy. If I could describe him in one word, it would be sweet. I've always felt that he and Sheila were good together. He appeared to truly love her and she was devoted to him. During the short time that I had known Sheila, she had shared with me some of their difficulties, even the fact that she had actually filed for divorce once. Fees had been paid and papers had been drawn up, but miraculously their hearts had also been changed. I hope that whatever is going on now, they will be able to overcome it as well. I've never been married, but secretly always wanted to have one like theirs. They weren't perfect, but they seem to be perfect for each other. It would hurt me deeply if anything ever happened to them.

Chapter Six
Tamara

I grew tired of waiting on Sheila to return, so I left for my lunch date with Samuel. I arrived at the restaurant first and was seated at a table. I looked up from scanning my menu just in time to see him walking through the entrance. *Lord, he is so handsome, and is that a Brook Brothers suit that he's wearing? Well at least the man has good taste and can dress himself.* I was staring so hard at his chest and broad shoulders that it wasn't until he had reached the table that I noticed he was holding a single long stemmed red rose.

"A beautiful rose for a beautiful woman," he announced as he handed it to me.

I reached for the rose, just as he bent down and kissed me on the cheek. My heart was racing and I instantly became hot. *Calm down, girl!*

"Thank you so much, Samuel. Are you always this considerate?" I questioned to cover up my delight.

"Only on special occasions," he answered.

"What makes today a special occasion?" I was curious.

"You agreeing to have lunch with me. I would call that special."

Oh, he's good!

"I have a good feeling about you, Tamara," he said.

"Do you really?" I asked as I fluttered my eyelashes at him.

"I noticed you even before you entered the coffee shop. I was struck by the confident way that you carried yourself—sophisticated with just a little sass. You're corporate when you need to be, but you can also bring out that girl from da hood. Versatile. I like that. From the first moment that I gazed into your lovely hazel eyes I told myself "I have got to know her name." I want to know everything there is to know about you. Your likes, your dislikes, what makes you smile and what makes you cry. I want to know all your dreams, so that I can make them come true. Because from this day forth I intend to be a part of your life, if you will let me," Samuel admitted.

I couldn't believe my ears. *Was this guy for real? All of this before we even had the chance to order lunch! Just my luck he's probably crazy, wanted by the police or married! He's just too good to be true!*

I made it back to work on a wing and a prayer, because my thoughts were completely absorbed of Samuel and our amazing lunch. I learned a lot about him in the two short hours we spent together and we had a lot in common. For starters he is originally from South Carolina, just like me. His father was in the military, just like mine. He's never been married, though he came close once. And has no children—at least, none that he knows of or is claiming on his taxes! He's successful, intelligent and owns his own home! I've struck gold! I must be living right, as my grandmother would always say.

It was a welcoming change to be in the company of a man such as Samuel. He genuinely seemed to be interested in the very essence of me. He wanted to know about my past, present and future. Not just what my preference was in bed, like all the other jerks I dated. Some didn't even beat around the bush about it. I guess they wanted to know if a second date was required. I always thought it came from bad rearing, but the longer I live I think it's built into the species as a whole. They seem to only think with their loins.

On those rare occasions when I was naive enough to let a man get close to me, he ended up disappointing me. Like the time I was madly in lust with Damon. I met him at the grocery store in front of the fruit section. He was thumping melons and I was grabbing at grapes. Our eyes met. We chit chatted about the outrageous prices for fruit. He then asked me for my phone number. He was cute, so I gave it to him. A few nights later, he called…finally. Our conversation was good. He asked if he could come by my place. I liked him, so I said "Yes". Next thing I know we were locking lips and touching each other places too personal for two people who had just met. He would come over just about every night and we would repeat the same hot action from the night before.

But after the dust settled, I started to wonder why he only came to my place and never invited me to his—or anywhere else for that matter. I questioned him about it. He came up with this bogus story about his condo was being renovated and the reason we never went anywhere was because he didn't want to share my attention with anyone or anything else. Sounded good at the time, so I bought it. But soon I became tired of him coming over to my place just to kiss and run. So after one of his little visits, I followed him—right up to the doorstep of his house furnished with a wife and kids! The smooth operator was married. I felt like a fool! I was so distraught over what I had uncovered that I didn't have the strength to confront him. I drove home and cried myself to sleep. I cried for what I had just discovered about Damon. I cried for the innocence my Uncle Ray took from me. And I cried for the realization that there was no Prince Charming in my future.

The next day I changed my number and my attitude. I was determined not to experience another broken heart, so I built a wall around my emotions. I kept them locked behind a metal door. It sometimes was a lonely world that I chose to live in, but at least it was a safe one. I wanted so badly to believe in love, but at a very young age that dream had been shattered. So, now I go through life numb. No love don't live here anymore, and it left no forwarding address.

When I arrived back to the office, I found Sheila was still playing her disappearing act. *Where in the world could she be?* By Leon's sudden visit I knew she wasn't with him. She's not doing anything relating to her Mama Lyn, seeing that she called looking for her too. It's true about not working with your family or friends because eventually they will start to take advantage of your kindness. Now Sheila's my girl, but I will still write her butt up if I had to. After all business is business. I must make a mental note to talk to sister girl about her afternoon disappearance. I need to nip this in the bud, before it gets out of hand.

The more I thought about Sheila's strange behavior, the more concerned I became. I had been back from lunch for over 45 minutes now and she still wasn't back. This is not like her. She is usually very dependable and rarely takes any time off at all. A thought came to mind. Clyde! I wondered if he was in his office or did he have to urgently leave out too. I wrestled with the idea of calling his office, but that would be prying and really it's none of my business. Then again, if she was with him on office time it was my business. I took it as long as I could. I had to call and find out. I dialed Clyde's extension. His secretary answered.

"Human Resources. How may I help you?"

I had to quickly think of a reason for calling.

"Good afternoon, Haley. This is Tamara in Benefits. Is Clyde in?"

"He is, but he's in a meeting right now. Would you like to leave him a message?"

Leaving a message was not an option. I had to know who he was meeting with.

"Well, it's imperative that I speak with him directly. It's regarding the number amounts for the new insurance packets. I found a problem." Which was a bold face lie!

"He shouldn't be much longer. As soon as he comes out of his meeting, I will make sure that he gets your message."

Wrong answer! I had come this far, too far to turn around now. I pressed on. "I'm sorry to be a pest, Haley, but this need to be handled right away. The new numbers are scheduled to go out this pay period. I will only be a minute."

She hesitated.

"Haley, are you still there?"

"Yes."

"May I ask who Clyde is meeting with that is so important he can't be interrupted?" I held my breath as I waited for her response.

"Well, he has been in a meeting with Sheila. He had me block off his afternoon, because he said that it was urgent that he met with her and he asked that they not be disturbed."

I felt like someone had just knocked the wind out of me!

"Do you know what they're meeting is about, Haley?"

"I was not privy to that information. I was just told to fit her on his schedule for today."

I was shocked, but didn't want to let on to Haley that she had just helped me solve Sheila's whereabouts. So, I made up yet another lie. "Oh, I remember now. I have been in so many meetings myself this morning that I had forgotten. I gave the numbers to Sheila. I'm sure that is what they are meeting about."

"Would you still like to leave a message?"

"No, that will not be necessary. As a matter of fact, there is no reason for you to even mention that I called." I quickly hung up the phone before I burst! To my horror, my intuition was right. Sheila was with Clyde! She had been hidden away with him all this time. While everyone was looking for her, she was deceitfully behind closed doors with the serpent himself, doing God knows what!

Chapter Seven
Sheila

From the moment I walked inside of Clyde's office I sensed my life would never be the same. He had the blinds closed with only one lamp glimmering from the corner of the room. I found him sitting on a sofa that decorated his office. He was sipping on something in a glass and had his suit jacket off. My first instinct was to turn around and walk away, but curiosity got the best of me, so I stayed.

"Sheila, please come in and make yourself comfortable. I tried to make the room less intimidating. I hope you don't mind," he smiled at me.

"No, actually it's very nice."

"Come sit next to me." He motioned for me to take a seat next to him.

I walked over to where he was sitting and settled in next to him, but not too close. I didn't want him to get the wrong idea. He leaned in closer to me and placed my hand in his. His eyes were all over me, especially my legs. I suddenly resented the slit in my skirt.

"You are looking very lovely today," Clyde commented licking his lips.

"Well, thank you for noticing," I said blushing.

"How could I not? You are a beautiful woman, Sheila."

"Thank you, Clyde, but you don't have to say that."

"Trust me, I wouldn't if it were not true," he stared into my eyes. I had to look away.

"I'm being rude. Can I offer you something to drink, sweetheart?"

"Yes, please."

"Coffee, tea or me?" Clyde joked.

"Water will be fine," I answered. I'm sure he was just innocently joking, but his flirting all of a sudden made me uneasy.

Clyde walked to the other side of the room and went into a mini refrigerator. He grabbed a bottle of water and poured it into a glass. As he handed me the glass, I noticed my hands were shaking. Clyde noticed it, too.

"Sheila, do I make you nervous? I promise not to bite, unless you want me to," he said seductively. He sat a little closer. We were close enough to kiss. I moved back. I wasn't quite ready for this. Clyde sensed my hesitation.

"Please forgive me for my forwardness, but I have been admiring you from afar for a long time. It's no secret that I find you desirable, but I'm not merely captivated by your beauty, it goes beyond that. I admire your strength and perseverance to do whatever it takes to keep your family intact. A woman who possesses such powerful qualities intimidates most men, but I find it to be a very attractive attribute. What man wouldn't need a wife who is going to have his back come what may. To me you are today's virtuous woman. If you were mine I would hold you in the highest esteem. I would shower you with affection and you would know without a shadow of a doubt that you are loved and appreciated."

As Clyde spoke to me, he was rubbing his hands up and down my leg. I wanted to stop him, but his touch felt so good. I watched his lips as he talked and wanted to taste them. He had me mesmerized. He continued rubbing up my leg and was almost at his destination, but I stopped him by putting my hand on top of his. He leaned into me and kissed me passionately, hard and long. I closed my eyes and enjoyed all the senses my body was experiencing. My body was on fire. At that moment I wanted Clyde. I wanted all of him and I wanted him to have all of me. I let Clyde kiss my neck, my ear and everything in between.

* * *

After everything had been said and done I shamefully fixed my clothes, left Clyde's office and had been hiding out in the ladies room ever since. I needed to pull myself together and regain my poise before I faced anyone. I didn't want anyone to look at me and be able to tell what I had just done. Me and Clyde, the Reverend Peterson. I was so ashamed.

I can't believe what I just let happen! What have I done? What has gotten into me? I just know I am going to burn in hell for this! Lord forgive me! How am I ever going to look Clyde in the face after this? How am I going to go home and look Leon, my husband, in the face? Lord, I'm sorry! I didn't mean for this to happen! I love my husband!

I never planned for this to happen. In the beginning Clyde was just a sounding board. I went to him to vent, get advice, some type of guidance. I was at the end of my rope. I had tried everything there was to try to help bring the fire back in my marriage. I planned get away weekends, secret rendezvous, wore sexy lingerie, prepared candlelight dinners, gave sensual massages, wrote love letters and read every couple's devotional book I could get my hands on. I suggested prayer and marriage counseling—all to no avail. Leon told me another man couldn't tell him nothing and he wasn't about to read a book. My hands were tied. I was a desperate woman in need of affection.

I loved Leon, but over the years I had fallen out of love with him. I didn't even look at him the same. He wasn't the man I fell in love with and married. He was so full of angry and I didn't know why. One minute he was as gentle as a lamb and the next he was an uncontrolled grizzly bear. His violent outbursts started soon after we were married. I could say one wrong thing and he would shout and scream like a madman. He didn't treat me as his loving wife anymore, but more like the enemy. He was easily offended and was quick to let me know. He had never struck me, but I wasn't so sure that he wouldn't. A woman shouldn't have to live in fear of her own husband.

Chapter Eight
Clyde

After Sheila had left his office, Clyde relaxed on the sofa while sipping from his glass of water and relishing on his most recent accomplishment…getting close to Sheila.

He had been watching her for the longest and had calculated his every move up until this very day. He was proud. *It took some time to seduce her, but she was worth the wait. Women are so weak and vulnerable. His strategy was to befriend them, tell them what they wanted to hear and then go in for the kill. It worked like* a *charm every time. When will they ever learn? Men aren't looking to be a woman's buddy. We have our male counterparts for that. We want one thing from them and one thing only. S-E-X!*

When you acquire their minds, they voluntarily give you their bodies. Talking to them as though you are actually interested in what they have to say helps to lower their guard. You build trust and then they start to reveal their deepest, darkest secrets. Secrets they only use to tell their girlfriends and before you know it, you are inside her head. She openly divulges her fears, her dreams, her disappointments and her failures; unconsciously letting you know in which direction you can then attack. I almost feel bad for them, not really. Foolish women, they should know better. They should learn to talk to their husbands or boyfriends and not an outside party. Only include a third party when the two involved agree upon it. Because if not, trouble will come lurking and sometimes even in the role of a pastor. Agreeing to be the pastor of Saint Luke Baptist Church is going to do wonders for my social life. Women flock to make sure that the man of God is taken care of.

Clyde laughed out loud at himself as he thought *I gots me a sweet scam goin' on here!*

Chapter Nine
Sheila

I just couldn't bring myself to go back to work. Not after what I did. I hid out in the ladies room in one of the stalls until I knew Tamara would be out of the office. Luckily, I kept a copy of her daily schedule, so I knew she had a 3:00 meeting today.

When I thought the coast was clear I made a mad dash to my desk, snatched my purse and high-tailed it out of there. All I wanted to do was crawl into bed and cover my shame. But before I could do that I had to pick the kids up from after-school care and get them situated. Tonight seemed as good as any to order a pizza for dinner and the kids would love it. I was in no mood to be over a hot stove. I had been in enough heat as it was. I tried to forget about this afternoon's events, but my mind kept drifting back to Clyde's office.

I had picked up the kids, the pizza and was pulling up in the driveway to find Leon's car already home. *What's he doing here at this time of day?*

"Daddy's home! Yeaaah!" Indya screamed as she ran into the house. Kelsey was close behind her. I followed with the pizza in tow. We found Leon in the kitchen preparing dinner. This day was beginning to be full of surprises. After he untangled himself free from Indya he came over to where I was standing and laid a wet kiss on me. I almost dropped the pizza onto the floor.

"Hey baby," Leon said.

"Well hello," I answered baffled at this man standing in my kitchen posing as my husband.

Leon grabbed the pizza from me and handed it to Kelsey. "Here son, you take Indya to the table and eat your pizza. Me and your mommy are going to have dinner in the dining room alone tonight."

"What brought all of this on?" I asked, still in disbelief.

"Can't a man cook for his woman? he said with a smile.

"Actually, I came by your office to take you out to lunch today."

"Today?" I interrupted him.

"Yes. Didn't Tamara tell you?"

"No, she didn't mention it," I told him.

"Well she said that you were in a meeting."

"Yeah, we both had meetings all day today," I lied.

"Well tonight we are celebrating you and all that you do. It hasn't gone unnoticed."

I couldn't believe my ears!

"Leon, I don't know what to say." My eyes were brimming with tears. I was both happy and ashamed.

"Just being my wife is enough." He grabbed me. "Now go upstairs. I already have a bubble bath waiting for you. Dinner will be ready by the time you get down." He kissed me on my forehead and just held me. "I'm sorry it took me so long." he whispered in my ear.

By now I am really bawling like crazy. While I was betraying my marriage vows with another man earlier today, my husband was preparing a romantic evening just for me.

I sat in the tub crying and pleading to the Lord for His forgiveness. I had asked the Lord to fix my situation, but I stepped ahead of him and helped myself. I only hoped I hadn't ruined things between Leon and me. I want nothing more but to grow old with him. *I have no other choice but to go to my grave with the secret I held. I have to do whatever it takes to keep Leon from finding out about what happened in Clyde's office today.*

I arrived in the dining room and there were candles everywhere, the table was set with our fine china and the food smelled delectable. I was feeling guiltier by the minute.

"Come sit here, my queen." Leon instructed me as he held a chair out for me to sit.

I followed his directions and sat down. He had prepared all my favorites. Grilled salmon, asparagus with a cream sauce, tossed salad and fresh, hot-baked crescent rolls.

"Leon, everything looks and smells delicious! I can't believe you did all this for me," I beamed.

"Of course, I did all of this for you. Who else would I do it for? Sheila, you are a very special woman and everything that I could ask for in a wife. I meant it when I called you my "queen", and for now on I am going to start treating you like one."

The tears started rolling again. "This really means a lot to me, Leon."

"And you mean a lot to me, baby. Let's make a toast," Leon suggested as he poured wine in our empty glasses.

"A toast," I said as I lifted my glass.

"To many more happy years together," Leon proclaimed.

"I'll drink to that."

"Now you're talking. Let's get this party started right. Eat up and by all means drink up."

That made me laugh. Leon could always bring a smile to my face. It had been so long that I have forgotten just how funny he could be. It was good to share a laugh with him again.

"Mmmm honey, this salmon is the bomb! What did you do to it?" I asked.

"Ancient Chinese secret," Leon answered.

"Boy, you ain't nowhere near Chinese! You don't even eat Chinese food!"

"Well it's still my secret!" He smiled.

"So you're not going to tell me?" I asked him seductively.

"It depends."

"Depends on what?"

"It depends on what you're willing to do to get it," Leon said with a sly grin.

"Whatever it takes," I said flirtatiously.

"Alright now. You keep talking like that and we are going to skip dinner and go straight to dessert!"

"That's fine by me!" I told him.

"Woman, I spent half of the day preparing this meal for you. Now your fast butt is going to sit here and eat it or get it fed to you!"

"Feed me big daddy!" I said.

We both laughed. For a brief moment, I had forgotten all about Clyde and this afternoon. We got through the rest of the meal, eating, laughing, talking and reminiscing. I hadn't thought much about the kids, until I spotted them peeking around the corner.

"Do you see what I see?" I asked Leon.

"Yeah, but I'm trying to ignore them."

"Mommy, can we come in? We want to have fun too?"

Indya asked.

"No, go away!" Leon joked.

Indya's eyes widened in surprise.

"Daddy's just playing with his baby. Come on in here and sit on daddy's lap."

Indya didn't waste anytime running into the room and jumping into Leon's lap. Then the phone rang.

"I'll get it," I said to Leon.

I guessed our little private party was over. I went into the kitchen to answer the phone.

"Hello?" When I heard the voice on the other end of the phone, it made this afternoon's incident come flooding back to my memory, confirming that the party was definitely over.

After the phone call I struggled to wash the dishes and put them away. I think subconsciously I was just stalling for time. Leon had the kids bathed and ready for bed. I had finally gotten in the bed and was scared to death of what Leon would want to do next. I waited for over 30 minutes for him to come to bed. What was taking him so long? He was supposed to be tucking Indya into bed and reading her a bedtime story. It usually doesn't take this long, but with Indya you'll never know. To avoid having to go to sleep she will have you read the same story twice sometimes three times. I tiptoed into the hallway and peered in Indya's room. I found Leon and Indya both asleep in her bed. *Thank God for small blessings!* I didn't attempt to wake him, but went back to bed…alone. After what I did this afternoon there was no way I could be intimate with Leon tonight. I'll just have to make it up to him another day.

Chapter Ten
Tamara

After my last meeting, I was tired and didn't even bother with going back into the office. It had been a long day and I was ready to go home. A hot bath and a glass of wine sounded like a good way to end my evening. I couldn't wait to get home and peel out of my clothes.

As I drove home I phoned in to check my voice messages from work. My run-away executive assistant had left me a message that she was taking the rest of the day off, and my potential new boyfriend had left one as well. How wonderful to know that he had been thinking about me. Especially, since I had been thinking about him all day myself. Between having him on my mind and Sheila, I couldn't even recall what my last few meetings were about. I called Sheila at home and see what her story was. Just when I was about to hang up she answered.

"Hey Sheila, I was just leaving the office and I got your message. I was calling to see if everything was alright with you."

"Oh yes, girl. Everything is fine now."

"Really," I said waiting for her to say more.

"Yes, thanks for calling to check on me. I'll be in the office bright and early in the morning."

"Okay. Oh, by the way Leon came by the office looking for you today." I told her trying to put her on the spot.

"Yes, he told me. We're just finishing dinner that Leon pleasantly surprised me with. So my bad day turned into a wonderful evening thanks to my loving husband."

Loving husband! She was really laying it on thick.

"Leon cooked? I bet you were pleasantly surprise. Well I don't want to hold you, so I'll let you get back to your 'loving husband'," I said teasing her.

"I'll see you in the morning and breakfast is on me!"

"I'm all for that! Tell Leon he needs to cook more often."

"Bye girl."

"Enjoy the rest of your evening and don't get pregnant tonight!" I joked as I hung up the line.

I guess all's well that ends well. Perhaps she was in Clyde's office today because she needed to confide in him regarding Leon. They obviously had a fight or some type of disagreement. I mean the man cooked dinner! That is not an everyday occurrence for Leon. He may be sweet, but from what I gather from Sheila, he is not the most thoughtful man. I felt bad thinking the worst concerning Clyde. Maybe he's not all that bad afterall. Now I will call Samuel back. I hope he hasn't left the office yet.

"Roberts' Accounting Services. May I help you?"

"Yes, is Samuel in?" I asked the receptionist.

"May I ask whose speaking?"

"Tamara Washington. I'm returning his call."

There was a dead silence.

"Hello?" I was slowly getting irritated. "Has he already left for the day?" As I spoke I could hear his voice in the background.

"Hold please," she told me.

What was her problem? I guess I wasn't the only one who had had a long day.

"This is Samuel."

My heart fluttered at merely the sound of his voice.

"Hey you. Sorry I missed your call. I've been in meetings all day."

"Well the wait was worth it and your timing was perfect. I was just about to call it a day. Are you headed home?" Samuel asked.

"I was."

"What are your plans once you get there?"

"Just a bath and a bottle of wine."

"Mind if I joined you?"

I was at a loss for words. Did he just ask me what I thought he asked me?

"I'm sorry, I meant for the bottle of wine—not the bath," Samuel corrected himself.

That set my mind at ease. I was beginning to think maybe I was mistaken about him and he was like the other boneheads I had dated.

"Sure I would love to see you. Why don't I give you a call with directions once I get home and get settled?"

"That would be fine. It will also give me a chance to

change from out of this monkey suit."

He gave me his home number and we hung up. I was really in a heated rush to get home now. I almost ran a stoplight trying to get to my destination. But after I arrived home I started to get second thoughts about him coming to my house. *Maybe we're taking things too fast. I only had lunch with this guy. Who's to say he's really who is trying to portray. What if he gets here and he's all over me or worse I'm all over him. It has been awhile and he was fine! I would hate to sabotage the relationship before it can start. Sex too soon definitely clouds the mind. I think I will call him back and just cancel. I would rather him be a little disappointed, then not have respect for me in the morning. Guess I will have a cold shower instead and kill the wine drinking. It only intensifies the mood.* I dialed Samuel's number. He answered on the first ring. "Samuel. This is Tamara."

"Tamara, that was quick. I just made it in the door."

"Well I called to see if we could reschedule for a later time."

"A later time? I was really looking forward to staring into those beautiful eyes of yours, but I guess I'll have to wait. You didn't get a better offer did you?"

"No, nothing like that. To be honest, I don't trust myself to be alone with you and I wouldn't want you to get the wrong idea about me."

Samuel laughed. "Tamara, I'm flattered. But I had no intentions in getting physical with you. Conversation is all that I'm after. So if it would make you feel better I will stay home and we can simply chat over the phone."

I was in heaven! *He's actually more interested in getting to know me as a total person than just sexually. I didn't know guys like him existed.* I poured myself a glass of wine and snuggled in for the night to converse with Samuel. It was midnight when I realized we were still on the phone. I went to bed with a smile on my face.

Chapter Eleven
Sheila

I arrived in the office bright and early with doughnuts in hand just as I promised Tamara. But I wasn't expecting to see what I did when I got there. On my desk were flowers and a card. They were tulips, my favorite. Leon had really outdone himself. First a romantic dinner and then this. I'm really going to plan something hot and juicy for him this weekend! I opened the card and almost fell to my knees when I read the contents. They weren't from Leon, but Clyde! I needed to trash this and fast, before Tamara arrived to the office. She would definitely want to know who they were from. I couldn't let her know they were from Clyde.

"Well, well, well. I don't know what Leon did, but after last night and these beautiful flowers I hope you have forgiven him." Tamara commented.

Tamara's voice startled me. It was too late to try and trash the card. So I quickly slid it back into the envelope, before she asked to read it. I hope she doesn't notice my nervousness.

"Oh, I have. He's done good." I smiled.

"I'm so happy to see you smiling again. I was really worried about you."

"That's sweet. You know me and Leon have our ups and downs, but things always have a way of turning around." I told her.

"I'm glad to know that because I love you both and hate to see you two at odds with each other. You are the closest thing I have to family, so you will have to excuse me when I get a little personal."

"I know Tamara and I appreciate it."

She stood there smiling at me as if she knew some big secret. She finally noticed the doughnuts I brought in.

"I hope this isn't what you brought in for breakfast! Lord have mercy. I thought I was going to come in to some cheese grits, bacon, eggs and hot homemade biscuits! This ain't hardly going to hold me until lunch!" she joked.

We both laughed. My ringing phone interrupted our laughter.

"I'll let you answer that. I'm gonna put my stuff down and

I'll be back shortly. We have some catching up to do," Tamara said.

Tamara walked off toward her office as I answered my phone.

"Good morning. Benefits Department. May I help you?"

"You most certainly may?" The voice on the other end of the phone said, dripping with seduction. Clyde.

"Hi there and how are you?" I asked.

"I'm doing very well, thank you. How are you?" Clyde responded.

"I'm fine."

"Well, I know that. But my question was how were you doing?" Clyde asked teasingly.

"You are just too much," I said tensely.

"So I've been told," he responded.

"Umm, thank you for the flowers. They're very nice."

"You're very welcome. I just wanted to let you know that I was thinking of you…still."

"That's nice to know."

"I can still smell you. When can I see you again?"

"Clyde, I really don't think that will be such a good idea," I whispered into the phone.

"Why not? Didn't you enjoy me? It seemed as though you did."

"Look Clyde, I'm married and you're married. So I really don't think we should see each other in that way again."

"Sheila, we both were married yesterday and that didn't seem to bother you then. So why are making it an issue today?"

Tamara walked out of her office and stood in front of my desk. I had to get Clyde off the phone and in a hurry!

"I will have to get back to you on that, sir. Give me some time to investigate and I will give you a call regarding my findings." I responded flatly.

"I guess by the change of topic, you are not alone."

"That's correct, sir."

"I'll be waiting for your call and please tell Tamara I said 'Hello'."

"I'll be sure to do that." I hung up the phone on that note.

What have I gotten myself into?

"Is there a problem? Tamara asked.

"No, I got it covered."

"Come into my office, so we can talk," Tamara said.

Oh my goodness. What now? I followed Tamara into her office racking my brain wondering what the devil she wanted to talk about. Before I could sit down she started.

"I didn't get a chance to tell you about my lunch date with Samuel!" Tamara was grinning from ear to ear.

"That's right. So how was it?" I asked relieved.

"It was amazing! He is amazing!" She was beaming.

"Really? Tell me more."

"We had lunch at Taylor Made Grill and talked for about two hours nonstop. We have so much in common. Girl, he's single and has no children. Now you know that alone makes him a hot commodity. Can you believe it? But that's not all! We talked later on that afternoon and he asked could he come over to my house. I explained to him that I wasn't comfortable with him coming over to my place so soon. He understood and suggested that we could talk on the phone until I was more comfortable. Sheila, we ended up talking on the phone until after midnight."

"I've yet to meet this guy, but he seems to be a keeper. Maybe the two of you should come over for dinner, so Leon and I can really size him up."

"Yes, Sheila! That would be great. I'm sure he would love to come and meet you all. I've already told him all about you. What about this weekend?"

"Pump the brakes. I've got my own surprise brewing for Mr. Leon this weekend."

"Such as?"

"Well first, I have to call Mama Lyn and see if she would keep the rug rats before I can set it off. Afterwards, I am going to visit Spencer's, my favorite adult pleasure store, and pick up us some new treats. You know I've been wanting a pair of those glow-in-the-dark dice."

"Alright now! Watch yourself!"

"Then I am going to decorate the house with candles, incense, rose petals and wine. Plenty of wine."

"And what about you? How are you going to be decorated?"

"Girl, I have this little sexy number that I've been saving and it's totally sheer. I'm gonna let it all hang out!" We both screamed!

"Leon ain't gonna know what hit him! If you need to leave a little early to get your weekend started go right ahead. It's Friday, so I just may slip out myself."

"Thanks, Tam. That will help me out a lot. You know things haven't been the best for Leon and I lately. But yesterday was like a turning point for us. So I want everything to be perfect tonight and throughout the weekend. Last night felt like we were newlyweds again. All the emotions that I felt for him during the early years came through and all it took was for him to show me that he still cared. The Lord is finally answering my prayers."

"Me and yours both. I think I have hit the jackpot with Samuel."

"When did you start praying, Ms. Thang?"

"After meeting Samuel. I had to give the man upstairs His props for creating such a fine specimen. I think I'm in love."

"Girl, you're a mess! I don't know what I'm gonna do with you."

* * *

After dodging Clyde's phone calls all day, I was finally on my way out of the office and on my way to Spencer's. They have everything your naughty little mind can imagine and more. Some of the stuff was a little too "out there" for even me and I was no prude. But that didn't stop me from coming out the freaky deaky store with an $85.00 charge to my Visa. Tonight was definitely going to be one to remember.

My next stop was the grocery store to pick up a few things. My last run was to the liquor store and then it was home to prepare my lover's paradise. I had talked with Mama Lyn and she was more than happy to keep the kids for the weekend. She even agreed to pick them up from school today.

After arriving home from the grocery store I worked

diligently on getting everything together. I had my fruit tray ready. The house was lit with scented candles and smelling good. I had rose petals leading to the bedroom with me sitting in the middle of our king-size sleigh bed. I suddenly become conscious of how tired I was from all of that running around today. I hope I don't fall asleep before Leon has a chance to get home. I jumped out of the bed and ran into the kitchen when I realized I had forgotten the wine in the freezer. Just as I was pouring the wine into the wine goblets I thought I heard Leon's car outside. Then I heard a car door slam. It was him. I struck a pose and waited for him to walk inside. By the look on his face I wasn't sure if he was happy to see me or not.

"Welcome home, baby," I purred.

"Wow! What is going on in here? Am I in the right house?"

"Of course, you are." I sashayed over to where he was standing, handed him a glass of wine and kissed him on the mouth. "I've been waiting to do that all day."

"Where are the kids?"

"You don't need to worry about them. They are taken care of. Now follow me upstairs, so that I can take care of you."

I took him by the hand and led him up the stairs. His mouth dropped open when he entered the bedroom. I positioned him at the foot of the bed, removed his shoes and started to massage his feet with lotion.

"Baby, can I shower first?"

"Just relax and let me do this."

After about ten minutes, I finished rubbing his feet with the massage lotion I had purchased from Spencer's. I poured him some more wine to help him to relax even further. I started to shower him with kisses, but before we could really get down to business the phone rang. But I choose to ignore it.

"Don't you think we should answer that?" Leon asked.

"No."

But Leon pushed me off and went to answer the phone anyway. I sat on the edge of the bed trying not to get pissed at the fact that he was more interested in who's on the other end of the phone, than me. He was talking rather low, which made it hard for me to hear the conversation due to the music I had playing. But I strained my ears to listen at any rate. He hung up before I could decipher anything.

"Honey, don't get upset. But I have to go."

"Go where?" I jumped off the bed.

"Me and the boys had planned a weekend fishing trip and they are waiting on me."

"What?"

"Yeah, remember I mentioned it to you the other night at dinner."

"You did no such thing."

"I could have sworn that I had told you about it honey."

"I believe I would have remembered something that big coming out of your mouth, Leon."

"It's no big deal."

"Leon, please tell me you are not about to walk out of this house."

"It's just for the weekend, Sheila. Just like when you go away on those weekend shopping trips with your church group."

"But I don't spring it on you at the last minute. It's planned months ahead of time.

Leon went to the closet and pulled out his already packed suitcase as I was talking to him. I became enraged. *So that was what that romantic dinner was all about the other night. It wasn't about us rekindling our love. It was about him trying to slip out of here for the weekend.*

"Leon, if you walk out of this house on me…"

"What, Sheila?"

"If you walk out of this house tonight, don't come back."

He picked up his suitcase and walked out of the bedroom. I heard the front door open and close again. I heard him starting up his car and I heard him driving off. I couldn't move. I stood in the same spot hoping it was all a bad joke and that he would walk back into the house. But he doesn't. He's gone. He left me for a weekend fishing trip with the guys…or was he even with "the guys" at all?

Chapter Twelve
Tamara

While Sheila was getting her weekend together I needed to make plans for my own. I called Samuel to see what his plans were. His response was whatever involved being with me. That man knew all the right things to say. I still wasn't comfortable with being totally alone with him. So we agreed to meet at Sugar & Spice for drinks. At night my favorite daytime coffee shop became an after hours café. I think I'll get a jump on him by leaving work early to give myself enough time to go home and change. With the warm weather we have been having my black and pink halter jumper would be perfect. Casual, but still sexy.

Interestingly, I was more excited about Sheila's weekend than my own. Just knowing that my suspicions about her and Clyde were false made my heart leap for joy. I also was feeling guilty for thinking such thoughts about Sheila. Clyde maybe, but Sheila never. Marriage meant too much to her. She's a class act all the way, not some heifer waiting in the shadows to jump another woman's husband. Cheating on Leon would be the last thing she would do. I shook my head for letting such outlandish thoughts enter my mind.

By the time I arrived at Sugar & Spice, Samuel was already there. He must have left work early too, because he was wearing jeans. I glided across the room with the most alluring stroll I could. The stiletto heels that I was wearing made it a difficult feat, but I gave it my best shot. Samuel-ever the gentleman-stood and pulled out my chair as I approached the table.

"You're looking lovely, as usual," he commented.

"Thank you very much, and I thought chivalry was dead."

"There's still a little lingering around. You just have to look in the right places."

"Obviously."

The waitress came over and took our order.

He asked me about my day and as I described it to him, he actually seemed to care about what I said. At times, if I said something that he didn't understand or was unfamiliar with he would stop me and inquire about it. I told him about Sheila and her weekend plans with Leon. I also mentioned her dinner invitation.

"You and Sheila appear more like girlfriends than co-workers."

"Yeah, we're pretty close. I love her like a sister."

"What did you tell her about me?"

"I just mentioned what a wonderful guy you are. That's about it."

"That's it, huh?"

"Yeah, that was all."

"Did you two already set a date for this dinner? I know she wants to check me out."

I looked surprised. "She's not going to 'check you out'. She only wants to meet you, because I had nothing but good things to say about you."

"And she wants to see if they are all true. Are you forgetting that I have sisters? I know how it's done," Samuel joked.

"Okay, you got me. I hope that doesn't mean you aren't interested in going?"

"Nonsense. I would love to go. Any friend of yours, I have got to make a friend of mine."

"Thanks. You're the best."

"Well we'll just see what Sheila has to say about that."

"Don't get it twisted. I do want Sheila to meet you and I would love for the both of you to become friends. But if that doesn't work out, it certainly doesn't change my mind about you."

"That's what's up."

The waitress brought our appetizers and the drinks we ordered.

"So what do you do for fun," I asked?

"I spend a lot of my free time with Jared. He's, my assistant, Melanie's son. Her husband's incarcerated. She is now a single parent of a five-year-old son, who is suddenly without a father. They were both devastated to say the least, to the point where her son stopped speaking. I felt for her, and even more for Jared, so I reached out to him. I started spending time with him, taking him fishing, playing basketball, football, and little league, all the things that a real father would do with a son. Slowly, he started to come around, and eventually started talking again. As a way of saying 'thanks' Melanie would have me over for dinner a few nights a week."

"Wow, that's amazing and so are you!"

"No, I'm just a guy trying to help out where I can. I enjoy doing things with Jared. Right now he needs me, and until his father comes back or Melanie tells me otherwise, I'm going to be there for him. There are just too many of our young black boys growing up without their fathers in the home. They don't have positive male role models in their lives of any kind except TI or 50 Cents. I don't want him to end up in the same situation that his father is in. It happens time and time again and that cycle has to be broken."

"You are really special Samuel. There needs to be more men like you."

"There needs to be more people willing to help one another. What's the purpose of living if you haven't made a difference in someone's life? I believe we are all here for each other. God blesses me, so that I can in turn bless someone else."

Did he say "God"? I almost choked on my drink. Now that's not a word I was accustomed to hearing coming from a guy I was dating—or any guy—unless it was followed with 'bless you' after I had sneezed.

"Tamara, I've been meaning to ask you a very important question."

"Yes, Samuel."

"Are you saved?"

"Saved?"

"Yes, saved. Have you accepted Jesus as your personal savior?"

I didn't know what to say. I've been baptized as a child. Did that make me saved?

Chapter Thirteen
Sheila

I had been in the bed crying the whole weekend. I tried calling Leon's cell phone a few times, but he wasn't answering it. He obviously was enjoying himself wherever he was. I was on my way to run some bath water, because by now I was sick of smelling myself, when the phone rang. I almost broke my toe running to answer it.

"Hello! Hello! Who? No, you must have the wrong number." I slammed the phone down in disappointment.

Even a stranger with the wrong number was able to get through and call the house, but not my wayward husband. I was feeling as low as a person could possibly feel. I needed some relief. I probably would have tried to make it to church, if I didn't look so bad. My eyes were so puffy from crying that I looked like a raccoon. So I soaked in the tub and cried some more.

"Lord why? Why does my marriage have to fall apart? Where did I go wrong? I don't understand. I've tried everything I knew to try. My heart is full of pain and I don't know if I can go on. Where are you? Why are you hiding yourself from me? I'm sorry about Clyde, I truly regret my sins. I asked for your forgiveness and I vowed it would never happen again. I'm tired, Lord. I need you now!"

I had to get myself together, before I headed over to Mama Lyn's house to pick the kids up. I could hide behind make-up, shades and a hat to cover my very nappy hair.

Two hours later I was parked in front of Mama Lyn's house. I took one last look in the mirror, before going in.

"Where is everybody?" I hollered as I entered the house.

"We're in here, baby," Mama Lyn answered from the kitchen. Indya ran out to greet me.

"Mommy! I missed you!" She hugged me so tight it almost brought me to tears again.

"Mommy missed you too, sweetie."

"Where's my daddy?"

I struggled to keep my composure. "You will see him soon enough, baby."

We walked into the kitchen together. Mama Lyn was

baking cookies and Kelsey was eating them as soon as they came out of the oven.

"Yo Ma. What it do shorty?"

"Hello to you too, son. Hi, Mama Lyn."

She turned her attention away from her cookies to speak to me. As soon as she looked into my face I could tell she knew something was wrong. The disguise didn't work. She knew me too well.

"Children, y'all gone and get your belongings together."

The kids left to gather their things just as Mama Lyn had told them to, which left us alone in the kitchen.

"Baby, what in the devil happened to you?"

I couldn't hold back any longer. The tears started rolling down my face and I opened my mouth to speak, but nothing would come out. Mama Lyn rushed over to where I was and wrapped her arms around me. This only made me cry even harder.

"Chile, whatever it is it's gonna be alright. Come sit down and calm yourself."

As I sat down at the kitchen table, Mama Lyn handed me a handkerchief.

"Here clean your face up, baby. Now whenever you're ready, you tell Mama Lyn what's got you all upset.

My lips quivered every time I attempted to explain. I would regain my composure only to lose it again.

"Just take your time, baby. I ain't going nowhere."

I wiped my tears and took a deep breath. "Leon left!" I blurted out.

"Whacha mean he left?"

"He's gone!"

"Gone where, chile?"

"When he got home Friday everything was going fine until he got a phone call. Afterwards he said he was going fishing with the guys. He said they were waiting on him and he just left! I told him if he left not to come back. But he left anyway. Mama Lyn, I haven't heard anything from him since. I tried calling him on his cell phone, but he won't answer. I don't know where he is or what he's doing, and I hate to think with who!" I collapsed in her arms.

"Hush, chile. Don't you fret. Everything is going to be

alright. Men lose their minds sometimes. But believe me, he will be back and he'll be happy to be there. You just take the kids home and wait til he get there."

"Wait until he gets there? Did you hear anything that I said? He left me! I don't know where he is or who he is with! I don't even know if I want him to come back home."

"Of course, you want him back. You talking foolish now, chile."

"I'm not talking foolish. He disrespected me and our marriage. I would have never walked out on him. I couldn't! Because he would have drug me back into the house by the roots of my hair!"

"But baby sometimes you just have to overlook some things, because that's how men are."

"Mama Lyn, I'm sorry but I don't agree with that. I feel a man should show the same respect they want shown to them. I've been a good wife to him and don't deserve to be treated like this. Now Mr. 'Gone Fishin' is going to get a piece of my mind when he does decide to walk his behind through the door."

"Now you hold your horses Miss Lady. Before you go giving people a piece of your mind you need to stop and think about your children. That's their father and they need him, just the same as you. You talking 'bout what you don't deserve. Do you think the Lord Jesus deserved to be crucified at the cross? No, He didn't. But He did it for us. Just as you have to endure some things for the betterment of the family. Things ain't gonna be good all the time. But in those vows that you recited before the Lord and all your friends you promised til death do you part! Not until he gets on your last nerves, not until the money run out, not until he messes up; but til death! Now that ain't nothing to take lightly." Mama Lyn paused to make sure I was paying attention, then she continued. "Yes, marriage is hard and it takes a lot of hard work. But anything worth having is worth fighting for. Some days you give and some days you take. But in the end it all balances out. You can't run out at the first sign of trouble. That's what's wrong with this new generation, too many options. Back in my day we stuck together. You better learn how to pray and praise Him until your change comes."

"Mama Lyn, I'm all prayed out. I don't even believe the Lord listens to me anymore."

"The devil is a lie! You got to stare Satan in the face and tell him to get out of your house! The victory is already yours, you just have to claim it! But you have to believe. Don't you believe chile?"

"I don't know anymore, Mama Lyn. I'm just tired of trying."

"But you are going about it wrong. You are trying to do it in your own strength. You have to get out of the way. Let go and let God. Chile, we share in His sufferings in order that we may also share in His glory. Don't give up in the middle of the race. You got to run on and see what the end is gonna to be. Do you hear me?"

"Yes ma'am."

"Now stop letting the devil get the best of you and get the best of him. Nothing happens in our lives that the Lord doesn't permit, and as long as I have lived on this earth I have never seen the righteous forsaken. You are more than a conqueror. Claim authority in Jesus name!"

"Okay."

"Now we are going to pray, and when you get home I want you to walk through every room and pray some more. Then when Leon walks through that door don't you fuss and just watch how the Lord beats him upside his head with guilt. I ain't telling you what I heard, but I'm telling you what I know to be true. Cause I have done it myself. You ain't got to punish him, the Lord will do it for you. Be the 'good wife'. Don't you try to change him. 'Cause you can't, but God can. Let the Lord have His way. Because His way is right."

Mama Lyn prayed a powerful prayer and sent me on my way. I felt renewed. I used the oil Mama Lyn gave me to pray and anoint each room in the house. *When Leon comes home he will find a different woman. Lord give me the strength!*

Chapter Fourteen
Tamara

Well that's the end of that. No more Samuel and me. I'm sure after him finding out over dinner the other night that I don't attend church he won't be interested in going out with me anymore. That was our "Last Supper". *How ironic. I knew he was too good to be true—too good for me anyway. What did a heathen such as myself look like dating a saint like Samuel? I'm not holy enough for him. He needed a sanctified woman. A woman with Jesus in her heart and a prayer on her lips. A woman who desires the things of the Lord, instead of the things of the world. Me and my fleshy ways would only distract him. He would probably miss heaven by simply being associated with me.*

I never get the good guys. It must be some sort of curse over me brought on by allowing my Uncle Ray to have his way with my body. *Is this God's way of punishing me for not fighting harder? If only I could go back in time and undo that tragic day. Maybe my life would be different. Maybe I would be different…better.* But that's only wishful thinking. I can't undo the past. So now my life is doomed with vile men who will never truly love me.

I wish I could have what Sheila has with Leon. I wish I had a man to look at me and see someone worthy. But the truth of the matter is I don't regard myself as worthy. Yes, I make a decent living. I had a nice home and wear designer clothes. I drive a fancy car. But, in all honesty, inside I feel my life is insignificant and without purpose. My existence doesn't matter. I could walk out my front door, not look back and no one would notice. No one would care.

It's really funny, other women look at me with envy. They notice all of the outside stuff and just assume that I have it all together. But what they don't know is that I cry myself to sleep many nights. They don't know that all my stuff can't replace the loneliness that haunts me. Because something inside was missing.

What makes Sheila smile in the midst of her trials? What motivates Samuel to do what he does for others? What keeps Mama Lyn going? I know they have a strong relationship with the Lord. *Is that the answer? Is that what is missing from my life? Would Jesus make the difference for me?*

Mama Lyn mentioned that she was praying for me. I wondered what her prayers consisted of. It's strange having someone tell me that they are praying for me when I don't even pray for myself. I figure what would be the use. I am sure God wouldn't hear me. I have lived my life so far from Him, He probably doesn't know my name anymore. Maybe He never did. In the midst of my thoughts I lazily reached over to answer the ringing phone.

"Hello."

"Good afternoon." It was Samuel! My whole mood changed and I do believe I was smiling.

"Hey you!"

"Were you busy?"

"No, I was actually just sitting around feeling sorry for myself."

"Now why would a beautiful, intelligent woman waste time feeling sorry for herself?"

"I thought I had lost your interest."

"What in the world would make you think something crazy like that?"

"To be frank, your religious beliefs and my lack thereof."

"Tamara, just because I love the Lord doesn't mean I don't have room in my heart for you too."

"But Samuel you are so good and righteous, I think I would be a bad influence on you."

"In the Bible it says that all things work together for good to them who loves Christ."

"Yeah and what does that have to do with me?"

"Because our meeting was no accident. It was predestined. I didn't intend on crashing into you. I'm sure you didn't wake up thinking some man was going to run into you with his coffee. But God did. He had already planned in time for our paths to cross at the precise moment that it did. Everything happens for a reason and God orchestrates it all. Even the bad events in our lives serve a purpose."

"I don't know if I can believe that. Why would God allow bad things to happen?" I questioned.

"What I'm saying is sometimes bad has to happen for good

to come through. Take Jesus being crucified. Now many thought in killing Him was the end, but in actuality it was only the beginning. Because through His death and resurrection it allowed for everyone after Him to be set free."

"What do you mean when you say 'set free'?"

"Saved. Those of us who have accepted the truth are rescued from the pits of hell. The devil no longer owns us. We belong to God and from that day of transformation we become a part of his 'royal family'. He looks out for us and protects us just as an earthly parent would their child. Now that's not to say we are exempt from heartache and pain. But it does mean we can live through it with the help of our Heavenly Father. All of our needs are met through Him and He promises never to leave us or forsake us. We are His to the end of time."

"'Royal family', huh? So how does one become a part of this 'royal family'?"

"The Good Book states that if you confess with your mouth and believe in your heart that God raised Jesus from the dead you are instantly a part of the family."

"And, that's it?"

"That's it!"

"So there's no running, jumping or shouting all over the place?"

"No, I think a lot of people think that you have to jump through hoops to get to salvation. But all you have to do is believe."

"Amazing. I don't think I have ever had it explained to me so simply. You have really enlightened me."

Because Samuel had already planned an afternoon with Jared our telephone conversation was brief, but his words floated around in my head. "God orchestrates it all" was what he said. *If this is true, what was the point in me being raped by my uncle as a child? What was the Lord attempting to do in my life with that incident? What good could come out of that?* None that I could see.

But I must admit not until now have I been interested in getting to know more about God or considered changing my life. I've never regarded anything or anyone past myself or my own accomplishments. It's always been all about me. I was all that mattered. After the many letdowns I had encountered I had become so disenchanted with life that a man and especially God could do nothing for me. But Samuel's attitude and love for God causes me to pause and reexamine myself. Could that be the reason for our paths crossing? Life Lesson # 2…For everything there is a season.

Chapter Fifteen
Sheila

I waited up for Leon until after midnight. I finally went to bed. But I left the table set, so that when he did decide to come home he would see the effort I made. I was so frustrated and my flesh kept rising up inside of me. I wanted to pack all of his stuff and set it outside on the curb and set it on fire. The nerve of him! Just who did he think he was staying gone all weekend without even a phone call! But before I could reach his dresser drawer and pull out a pair of his boxers, Mama Lyn's words echoed in my head. I had to fight. The devil wants me to give up and throw in the towel, but I owe it to God to fight. But I don't know if I had anymore fight left in me. I've been trying since day one to make this marriage work and I'm now running on empty. It would be a welcoming change to have a little help. As I drifted off to sleep I pleaded to the Lord to breathe new life into my dying marriage.

I'm not sure what time it was, but I was awakened by a noise coming from downstairs. Out of habit I reached over to Leon's side of the bed to wake him, but he still wasn't there. I was both fearful and angry, because of the predicament I was now in. But I had no time to waste on self-pity, a stranger was in my house. So I slipped out of the bed grabbing my curling iron in one hand and a heavy can of hairspray in the other. I guess when I got downstairs I was going to style the intruder's hair. I slowly crept down the stairs. There was a lot of noise coming from of the kitchen. What was he doing? Fixing a snack, before he robbed me blind? Just as I turned the corner, he did too and we bumped into each other. I screamed, but not before I sprayed hairspray in his eyes. Now it was his turn to scream.

"Leon! What are you doing? You scared me half to death!"

"What the hell are you doing spraying me with pepper spray?"

"I'm sorry and it's not pepper spray, its hairspray. I heard noises and I thought someone had broken into the house. So I snatched the first thing I could get my hands on. Afterall, I was home alone and was only trying to protect myself and the kids," I said, following behind him into the downstairs bathroom as he rinsed his eyes in the sink. "Let me help you."

"No thanks!" he snapped. "You've done enough!"

"I didn't do it on purpose, Leon."

"I don't know that! You may have," he yelled.

"Don't be ridiculous!" I hollered back. He stood up and dried his face off with one of the good towels I hung in the bathroom for decoration. But I didn't think now was the time to bring that up.

"I'm not being ridiculous. I'm being realistic."

"Realistic? How do you figure that?"

"This is probably your way of getting back at me for leaving for the weekend. You women can be so devious!" He pushed past me.

"You women! I'm not some woman, I'm your wife and you should know me better than that." I said to him fast on his tracks. "Have you even stopped long enough to look around down here? Have you noticed the table setting, the dinner or the candles? While you were rambling around in the kitchen did you notice the unopened bottle of wine in the refrigerator? I went to all that trouble for you!"

"Well, I didn't ask you to go into any trouble for me."

"Leon, I'm sorry. It wasn't any trouble and you never have to ask. I do it, because I love you. Leon, please I don't want to argue."

"Who's arguing? Not me. I'm tired. It's been a long weekend and I'm going to bed."

"Leon, don't walk out on me again!"

"Sheila, please no dramatics tonight."

With that he went upstairs and had the nerve to close the bedroom door. Before he could make it to the top of the stairs I collapsed on the floor. I fell face down onto the floor and let go of all of the frustration, disappointment and pain that I was feeling. I lay in a fetal position, hugged my knees and sobbed uncontrollably. *Why is he treating me this way? Haven't I been a good wife?* I value marriage and the significance of each partner's role. I understood the importance of making your mate feel appreciated even when the world may think otherwise.

Apparently, Leon didn't share the same perspective. Either

that or someone else had his attention. I had wondered in the back of my mind for some time if there was another woman in the picture. But just the mere thought of it was too much for me to fathom. So I discarded the notion as quickly as it came. Eventually, I had to face the truth. Especially after this weekend when he blatantly showed me he wasn't interested in me. He was a man, afterall, so he had to be interested in someone. With that revelation I pleaded to the Lord with yet another request.

"Father, I humbly come before you broken and in despair. My spirit is heavy with grief and I cry out to you for comfort. Only you can lift my burdens. I am on the verge of giving up all hope. But before I do, please intervene. The man I call my husband has become a stranger to me. I don't know him anymore. I've done all I know to do to keep this marriage together, but it's still not enough. Lord, I beg you to step in. I ask you to remove any outside influences that are interfering in our marriage. Anything or anyone who is blocking our growth as a couple, I rebuke it right now in the name of Jesus. Whatever tactic the enemy is using to separate us I ask that it be revealed and destroyed. Cover us with your arm of protection. Keep us Father, when we are unable to keep ourselves. Amen."

I woke up when the sun shining in from the dining room window hit me in the face. I had fallen asleep on the floor. I really must have been tired to have slept on that hard floor all night. I can't believe Leon let me stay down here. He really wasn't concerned about me anymore. But I wasn't going down that road today. I had to remain focused.

I went into the bathroom downstairs and washed my face. I then went into the kitchen to put a pot of coffee on. I turned on my George Foreman grill to fry some bacon when I heard the shower going upstairs. That meant Leon was up. *I'll take him a fresh cup of coffee. Lord help me.*

As I was entering the bedroom, he was coming out of the bathroom still dripping. In spite of everything, he could still make me melt. I handed him his cup of coffee with a smile.

"Good morning, Leon."

"Morning." he reached for the cup and brushed past me.

"Did you sleep well?"

"Ummhmmm."

"I'm glad. I was going to fry some bacon. Would you like

some?"

"No thanks."

"Can I fix you a lunch for work then?"

"No time."

The whole time I am talking to him he was steadily getting dressed. Not once does he look up or acknowledge me. I sat on the bed next to him. He jumped up, grabbed his keys and headed for the door. But just before he exited he turned and said "Thank you for the coffee." *Thank you for the coffee! That's all I get! Not even a kiss goodbye. I'm starting to feel like a real fool. When is enough, enough?*

Chapter Sixteen
Tamara

I couldn't wait until I got into the office and hear all about Sheila's hot weekend. Sheila is the queen of romance, so I knew it had to be out of this world. There aren't too many women who will go all out to make their man feel special after so many years of marriage. Sheila is one in a million in my book. I stopped into my favorite coffee shop and picked us up a few goodies. I thought we could snack on some treats as she gave me all the highlights of her weekend.

I bumped into one of the other directors on the elevator. There were only a few women who held the position as director, and even fewer who were Hispanic. Monique Rodriquez was in her early forties and recently divorced. She was a short firecracker, quick on her feet and even quicker with her tongue. She wasn't one to hold back how she felt at any hour of the day. We had often been present in the same meetings and she went toe to toe with our male colleagues. I have always liked Monique, mainly, because you always knew where you stood with her.

"Good morning, Monique."

"Morning girl. I'm not sure how good it is yet. It's too early to tell," Monique responded.

"How was your weekend?" I asked her.

"Too doggoned short! Whose bright idea was it to work five days and be off two anyway? I'll like to meet them and choke them with my bare hands."

Like I said, I've always liked Monique. You could count on her to always keep it real. We arrived on her floor first. The doors opened and she proceeded to walk off the elevator. Then she stopped, turned around and pointed at the bag I was holding.

"What's in the bag?" She asked.

"Cheese Danishes. Would you like one?"

But before I could get the words out, she was already reaching for the bag.

"I don't mind, if I do. I didn't have time to eat anything with me trying to get my kids dressed and ready for school. Not that I had the money to get something anyway, since my good for nothing ex-husband is late with my child support again. But I ain't complaining, I'm just explaining." She said it all in one long breath.

The rest of the elevator ride to my floor was quiet. I was alone with my thoughts and I suddenly felt sorry for Monique. It came to me why she appeared so harsh. Her life was harsh. I didn't have any kids and couldn't imagine trying to raise them on my own. I have a hard enough time just dealing with myself.

I finally made it to my floor. I dashed off the elevator and made a beeline for Sheila's desk. But to my disappointment she wasn't there. This was strange, because she is usually the first one in. Promptness was her middle name. I, on the other hand, was always late. *I guess her weekend was even better than I thought.* Now I was really excited and couldn't wait to hear every juicy, tantalizing detail. But I would just have to settle for checking my e-mail, until Sheila made her way into the office.

Halfway, through checking my messages it dawned on me that Sheila still hadn't come in. I was starting to worry and was just about to pick up the phone to call her house, when I heard a noise outside my office door. It was Sheila and she looked like death warmed over.

"Sheila!" Her appearance startled me.

"I'm sorry I'm late. I had a bad morning. I would have called, but then I thought I would make it on time. But then I got stuck in traffic and …"

"Sheila. Stop! What's wrong?" She was just rambling. I walked over to her and put my hand on her shoulder. "Sheila, talk to me."

She looked up at me and her eyelids were brimming with tears. One rolled down her cheek and then it was like the floodgates opened. Tears were flowing, but she wasn't making a sound.

"Oh my goodness, Sheila. What happened?" I grabbed her by her hand and lead her into my office.

"Have a seat." I handed her some tissues. I sat quietly by her side. Every time she opened her mouth to say something, her lips trembled but nothing came out.

"Take your time, Sheila," I rubbed her back as I spoke to her. I was trying to sooth her, but my insides were a mess! With each second I thought I would explode.

"I don't think I can do it any longer," Sheila finally said barely above a whisper.

"Do what?" I asked her.

"Stay married to Leon."

I almost lost my breath. I couldn't believe what I was hearing.

"What? Sheila, please things can't be that bad. What happened? What in the world has Leon done this time? What did he do get himself a girlfriend?"

I was talking so fast and was asking so many questions at once that I didn't realize what I was saying. She didn't say anything. She held her head down and cried.

"No!" I wanted to cry with her, but I knew I had to be strong. "How did you find out?"

"He was gone the whole weekend!" she responded.

"Gone where?"

"He told me he was going fishing with the guys?" she answered angrily.

"He left you to go fishing?" Now I was angry! Who did Leon think was going to believe that bullcrap story? I certainly wasn't. But I didn't want to fuel Sheila's fire, so I kept my thoughts to myself.

"Tamara, I've done everything and Leon has done nothing to make this marriage work. I'm tired and I can't take much more. I am seriously thinking about leaving him."

I didn't know what to say. I was wracking my brain trying to come up with something wise and profound, when I heard Monique's voice coming from out of nowhere.

"Think smart!" Monique said leaning inside my office doorway.

We both were caught by surprise and were speechless.

"I didn't mean to eavesdrop. Tamara, I actually came by to tell you thank you for the Danish. I wasn't being rude on purpose this morning, but I had a lot on my mind," she said as she made her way over to where we were sitting. "Sheila, if you don't mind me saying, I've been where you are and I know it's not easy. But divorce is no cakewalk either. It's a death. A death of a marriage, a family, and your life as you know it. So before you take that step to the other side think it through."

"Are you saying you regret getting your divorce, Monique?" Sheila asked.

"Hell no! I only want you to know that divorce is no easy feat, especially with children involved. Their lives are also being turned upside down and inside out. They don't understand that mommy and daddy aren't getting along. They just want the two of you living together under the same roof. So don't think about how you feel right now or how tired you are, think about the damage it will do to your kids."

"Did you think of the damage it would do to your kids?" Sheila asked sarcastically.

"In my case, it would have been more damaging to stay in the household with that crazy nut! I have no regrets and I want you to be sure that you don't. Sometimes our emotions will make us fly off the handle and think irrationally. So slow down and think smart. Go talk with an attorney, if you think that would help. Talk to your pastor. Exhaust all of your resources because after the papers have been signed, there's no turning back."

"I honestly don't want a divorce. I'm just so mad and hurt by the way Leon's been acting towards me. I thought this last resort would get his attention," Sheila admitted as she exhaled.

"Well I suggest you scream, holler, curse him out, or jump up and down. But don't run and get a divorce! That's not the answer. Have you even asked him what he wanted?" Monique asked Sheila.

"No, I guess I haven't."

"Then that's where you need to start. Stop doing things trying to get a response out of him and just be straight up. Go home and ask your husband does he still want to be married to you. If he says "Yes", then tell him some things need to change. Talk to him and tell him how you feel. Don't assume that he knows. Men aren't that bright. Haven't you realized that by now? Well I have given enough of unwanted advice. I have work to do. Thank you for the Danish, Tamara." Monique was talking as she walked toward the door.

I was grateful that Monique and I took the same elevator

this morning—and for her fondness of cheese Danishes.

Chapter Seventeen
Leon

I didn't mean to treat Sheila so harshly, but here lately she has really frustrated me. There was a time when we would have so much fun together. We talked about everything and shared many laughs in between. She was my best friend and I was her hero. Whenever she looked at me, there was a sparkle in her eyes. But now the light has gone out and disappointment and darkness are all that is left. She no longer sees me as her hero, and that devastates me. I would do almost anything to bring that same light back in her eyes. But nothing I do pleases Sheila. Nothing is ever good enough. She wants things her way and anything different from that is not acceptable. Everything has to be on her terms. Our marriage, our children, our home, even the way I dressed had to be to her liking. Anytime I had an opinion or suggestion it was not well received. So I wondered why I am here at all. I wasn't needed or respected. She may as well be married to herself. She doesn't want a husband, she wants a clone.

At times when I felt I couldn't take much more, I turned to my buddy and co-worker, Al. Al had been married fifteen years to the same woman. They have their share of problems, like the rest of us, but he truly loves his wife and she respects him. He once confided in me that in all of their fifteen years together that he had never once cheated on her.

I looked up to Al. He was the closest man I had to a father figure. My own father wasn't around much while I was growing up. He was in and out of the house, sometimes staying away for months at a time. My mother finally got tired of him and all of his other women and kicked him out of the house for good. He left without even a good-bye. I was in the seventh grade at the time and to this day I have never forgiven him. He chose his women over me and left me to raise myself. I had to become a man without any formal training. I made a lot of mistakes along the way including my first marriage, which ended in divorce, because I wasn't equipped for the job. All of this I blamed on my absentee father.

I wanted to be and do more for my son and daughter. I

never wanted them to feel rejected or worthless, cast-out or thrown away. Those were the feelings that haunted me throughout my childhood after my father left. I was convinced that something was wrong with me. I often told myself if I had been important or special, he would have stayed. The same demons tormented me today. They surfaced every time Sheila finds errors in the things I do. She makes me feel like that worthless little boy my father left and my only defense was to fight. It was all I knew.

I didn't know how to tell her about my inadequacies, so I talked to Al. Sometimes only a man can understand another man. I'm sure she thought the worst when I went off for the weekend, but the truth is that I spent it with Al. He invited me weeks ago to come along with him and his brother-in-law to a men's retreat. I lied to Sheila about the fishing part because I was too embarrassed to tell her the truth. How do you tell your wife that you need help being a man so that you can be a better husband and father? I realize I have some shortcomings, but I don't want to hear Sheila play them back to me on a continuous basis. She criticizes me enough on her own. I hated to keep this from her. Particularly, since we used to share everything. I was between a rock and a hard place. But I don't think she would understand my struggle especially since she thinks she is near perfect and doesn't have any struggles of her own.

Chapter Eighteen
Tamara

I had been so engrossed in Sheila's issues that it was lunchtime before I realized I hadn't spoken with my own man. I hadn't heard from Samuel, which was unusual. Since he hadn't called me, I decided to phone his office. I was so grateful to have him in my life, especially with my history of duds. Men like Samuel were rare. He was so full of confidence and self-assurance.

I used to sit and wonder how a man could jump from woman to woman without any commitment or why he would degrade her during the day, but lay down with her at night. It was simple. He simply didn't value himself. So therefore he couldn't value anyone else.

This insight came to me after one of me and Samuel's many conversations. He stated the Bible says that a man should love his wife as his own body. Now, who would do harm to his own body, unless he first did not value it? So many men lacked the knowledge of true love. Unfortunately, so many women were lacking too. I had been one of them.

After calling Samuel's office his secretary informed me that he was unavailable. But before I could leave him a message she rudely hung up the phone. Now that pissed me off more than a little bit. Mainly, because I knew that hateful witch did it on purpose. So me being the woman that I am, I called right back!

"Hello. Roberts' Accounting Services. How may I help you?"

"Yes, Melanie, this is Ms. Washington." I knew that with me addressing her by her first name would bring her down a peg or two. "You disconnected the line before I was able to leave Samuel a message, dear."

"And what might that be?"

Sister girl had an attitude! I liked that because I have one, too.

"Tell my sweetie that I called and to please give me a call back. I haven't heard from him all morning and I miss hearing his voice," I snickered.

"I'll be sure to let him know that you called…again, but you can relay the other half of that pitiful message yourself. This is a place of business, not a dating service Ms. Washington."

With that she hung up the phone in my face…again! *Oh, no she just didn't!* I grabbed my purse and headed to Cherry Street, where Samuel's office was located. What I wanted to say to her needed to be said live and in person. She obviously didn't know who she was messing with, but she was about to find out.

I couldn't get to Samuel's office fast enough. As soon as I entered I spotted my adversary. It had to be her. She was the only one sitting in the receptionist's area. I stopped in my tracks. She was attractive; I might even go so far as to call her beautiful. Though she was sitting, she appeared to be tall in stature with a cocoa-brown complexion and long flowing hair, (probably a weave). She looked up from what she was doing and our eyes met. It was too late to turn around and bolt out of there. *I had to finish what I had come to do: Put that witch in her place!* I walked towards her, our eyes never leaving one another.

"Melanie, I presume."

"You presume correctly," she answered eyeballing me over from head to toe.

"I'm Tamara Washington," I proudly announced to her. She didn't even flinch. It was as though she already knew who I was and was expecting me. "I thought it best that we met face to face."

"Oh, really and why is that?" she asked icily.

"I felt you needed to know that I am now a special part of Samuel's life and have no intentions on going anywhere. You had best get acclimated to the idea of me being around. So drop the bad attitude. Because whether you like it or not, I am the head woman in charge. So respect my position or I will see to it that you are standing in the unemployment line." I stood there really full of myself waiting for her to recover from my tongue-lashing and give me a response, when I heard Samuel's voice behind me.

"Tamara!" He barked. My body stiffened.

I spun around to face Samuel and his fury. I was both petrified and embarrassed: I feared losing Samuel, and I was also ashamed of my behavior.

"Samuel, I…"

"I think we need to speak in my office."

I reluctantly followed him to his office. Once inside he closed the door.

"What was that display all about?"

"Hear me out, Samuel. It was Melanie. I grew tired of her disrespect towards me."

"So you thought it wise to come to my place of business and cause a scene? I thought you were better than that Tamara."

"I know I let my emotions get the best of me and overacted, but Melanie was rude and…"

"My employee, which also makes her my responsibility. If you had a problem with her you should have brought it to me and allowed me to deal with it, instead of causing a scene and undermining my authority. Now I think you owe her an apology."

I was starting to get a little miffed with him. *He doesn't even know how everything went down. He only walked in toward the end of our conversation. He didn't hear the way in which she had talked to me. Now I am looking like the bad guy.*

I was really in a dilemma. I looked from Samuel to the door. Either I save face and apologize or walk away from the most fulfilling relationship I could ever have with my pride still intact. *How is Melanie ever going to respect me if I bow down to her now?*

"I apologize for losing my composure and causing a commotion, but I will not be apologizing to 'your employee.' Melanie was very unprofessional and obviously has a problem with me. You need to open up your eyes and see her for who she really is."

With that I walked out of the office and onto the busy street, hoping Samuel would follow after me. He didn't. It was a long ride back to my office and I had an unsettling feeling that my relationship with Samuel had changed for the worse that afternoon.

Chapter Nineteen
Sheila

I was sitting at my desk meditating on some of my favorite scriptures from out of the book of Psalms. I needed to get my mind and spirit centered. In all the years that Leon and I have known each other this had to be the scariest time in our lives, at least it was for me. Not knowing if we were going to make it was a frightening concept. Being mad and not speaking for a few days was totally different than not living together as husband and wife. The realization of it all suddenly hit me hard. Leon had been my first love. Most of my firsts were experienced with him. He was my first official boyfriend, my first true love, the first man I cried with, laughed with, made a home with, worshipped with, partied with, dreamed with and prayed with. My life seemed to have not started until after our paths crossed.

I was heavily into my thoughts and had not noticed Clyde, until it was too late. He zapped me back to reality with a soft kiss on my cheek. I jumped.

"I'm sorry, I didn't mean to startle you sweetness."

I was more than startled—I was livid!

"Clyde, please don't do that. Don't ever do that again!" I just stared up at him, dumbfounded by his boldness. I wasn't sure what his problem was, but he was getting a bit out of hand and a little too comfortable.

"Sheila, why are you turning my very innocent gesture into something more than how it was meant? I am here only out of concern. I didn't see you in church yesterday and I came by to check on you and make sure everything was okay. I realize we have a work relationship, but I'm now also your pastor and I was ensuring all was well with you. I'm sorry for the confusion. Have a blessed day." He started to walk away.

Now I felt bad for practically biting Clyde's head off. Clyde has always been a friend to me. The person I could always talk to. Now I was treating him like the enemy.

"Clyde wait! I'm sorry. I've had a terrible weekend and I didn't mean to take it out on you."

He turned to face me with a look of alarm. "Is there

anything I can do?" he asked with what appeared to be genuine interest.

"Just pray. I need a blessing and I need it right away."

"I pray for you each day, Sheila. I know you've stated you wanted nothing more to do with me, but I still care for you and whatever you're going through please know that I'm here for you always. I also suggest you read Psalm 91 in its entirety. Be blessed."

This time he left and didn't return. How uncanny that Psalm 91 was one of the scriptures that I had just meditated on before he had arrived. Maybe that was a sign from the Lord that Clyde was a friend indeed.

Chapter Twenty
Tamara

Now I'm really starting to worry. It's Tuesday and I still haven't heard from Samuel. We usually talk every evening until one of us falls asleep on the phone—usually me. Sheila was still looking down in the dumps. I supposed her talk with Leon didn't go well. But I didn't have the strength to ask her about it. I was consumed with my own thoughts of Samuel and it was slowly driving me insane. Pretty soon me and Sheila are both going to resemble the walking dead. Why is it that women always lose their minds over a man? We lose ourselves in them and before long we don't exist anymore. We're merely his wife or his girlfriend, with no identity of our own. Well, I'm not going down like that. I needed a diversion. Food!

"Hey Sheila girl, you want to walk down to the cafeteria with me and grab an afternoon snack?"

"No, you go ahead. I don't have much of an appetite."

"Well, I just asked you to be nice. You really don't have a choice. So get up from there and let's walk."

"Tamara, I am not in the mood."

"Girl puleeze! Every woman of every age is always in the mood for something sweet, particularly chocolate, and especially when she is having a bad day. It's one of those unwritten rules. Now come on. Let's go drown our miseries in something sinfully high in calories. I'm not moving from this spot until you agree to go with me." I said as I sat in one of the chairs in her office.

Sheila saved her work on the computer and stood up. "Well since you are not going to allow me to get any work done, let's go."

The cafeteria had an assortment of muffins, Danishes, cakes, pies, cookies and brownies. There was also fresh fruit, but we bypassed all the healthy stuff and went straight to the sweets. I helped myself to the brownies. Sheila purchased the peanut butter cookies. She was going to have to share some of those with me. We found a booth and went to work on our sugary treats.

"Mmm…heaven." I commented after biting into my brownie.

"Uh, uh girl. You have got to taste these cookies!" Sheila said.

"I thought you would never ask." I broke off a piece of her cookie and popped it into my mouth. "Mmm…now that's divine."

"Almost better than sex," Sheila boasted.

"Almost!" We both laughed.

"Now you know I have to ask. How are things with you and Leon?"

"Girl, I don't want to talk about me and Leon. I'm trying to enjoy my cookies. So don't ruin my sugar high. What about you and Mr. Perfect?"

"Of course, you know me I've already mess up things between us. He didn't even call me last night."

"Oh my goodness. What did you do?"

"He walked in as I was giving his secretary a piece of my mind."

"Girl, I'm scared to ask. But what did you say to her and why?"

"Sheila, believe me when I say this sista needed to be checked."

"Ummhmm…just spill the beans."

"First of all, she is rude. From day one she has had a problem with me. Whenever I call she always gives me attitude. How is this for unprofessional, I called yesterday and she hung the phone up in my face. I could have been a client and she's treating me like I stole something. After that I felt I needed to pay sista girl a visit, so I went down there."

"To the office?"

"No, to her house. Of course, the office!"

"Well you don't have to get funky with me. I'm not the one you are mad with."

"See how she affects me?"

"I see how you are allowing her to affect you."

"You don't understand Sheila. I know she is trying to sabotage me and Samuel's relationship. I can't let that happen. He is the best thing that has happen to me. The worst part is Samuel said that I needed to apologize."

"So you think by going down to Samuel's office, his place

of employment and in your own words 'check' his secretary was a smart move? I'm with Samuel. You do need to apologize."

"Oh, I won't be apologizing!"

"Tamara, sometimes you have to lay pride aside and be the bigger person. Kill her with kindness."

"But she was wrong."

"Look at it from Samuel's point of view. You wouldn't have him coming up in our office and acting a fool. Not calling you a fool, but you did play yourself."

After our sweets excursion and much consideration, I decided to make another trip downtown. I had a secretary to apologize to.

I walked into the office to find Melanie in the same place as she was before. But this time she was on the phone and didn't notice when I walked in. I waited patiently as she finished her conversation. I took a deep breath, put on my widest smile, and approached with caution.

"Good afternoon, Melanie. I came by to apologize for my very inappropriate behavior on yesterday. I was out of line."

"I'm sure Mr. Roberts, or 'Samuel; as you so fondly call him, will be happy to hear that. But your little, weak, day-late apology doesn't mean anything to me."

"Melanie, this isn't for Samuel. I truly mean it. I feel ashamed of myself for acting the way I did. I had no right."

"Who do you think you're talking to? Melanie leaned back in her seat with her arms folded. "You and I both know if Samuel would not have witnessed your childish temper tantrum, you wouldn't have thought twice about apologizing."

I could see I was getting nowhere with her and decided to back off.

"Once again, I'm very sorry. I hope you can find it in yourself to forgive me for my rudeness. Have a good day." I turned to leave before she could respond. *That certainly didn't go as well as I expected. Thank goodness eating crow is not one of the necessary food groups.*

Chapter Twenty-One
Sheila

Just before I was about to leave for the day, Clyde called. He wanted to know if I was feeling better. I told him after our conversation that I was hopeful. It put my heart at ease to know that he was concerned and was praying for me and Leon. I thanked him for being there for me…again.

It was a relief that we could put that one sordid incident aside and still remain friends. I still can't believe I allowed my flesh to get the best of me. But I now realized that I'm not beyond temptation and that no matter how saved you think you are, you still have to pray without ceasing.

When I arrived home, my plan was to cook dinner and get the kids in the bed as quickly as possible. I wanted to have a heart-to-heart with Leon without any interruptions. Getting the kids to bed was no easy task. Indya cried for her daddy and Kelsey couldn't understand why I wouldn't let him play a few more games on his Play Station 3. But with much coaxing and a bribe of five dollars a piece, they finally went willingly.

Now I was alone with my thoughts. I paced the floor and rehearsed in my mind the words I would say to Leon. Outside I heard Leon drive up. This was it! I silently said a prayer. I was seated at the dining room table when he came into the house.

"Hey," Leon announced his presence.

"Hi Leon. How was your day?" I asked. I could smell the oil from the engines he had worked on at the shop and it was visible on the overalls he was wearing.

"Same ole, same ole," he answered.

"Are you hungry? I have a plate for you in the microwave."

"No thanks, I don't want anything," he said walking upstairs.

"Leon, could I please talk to you before you go upstairs?"

He stopped and turned around.

"Come have a seat here with me." I patted the chair next to me.

He slowly walked over and took a seat. "What's this all about, Sheila? I'm really tired and don't feel like getting into it tonight."

"I only wanted to talk with you."

"About?"

"Us."

"What about us?" he sighed.

"Well, things haven't been the best between us lately and I wanted to talk about it."

"What's there to talk about?"

"Well, Leon things can't go on like this. I wanted to discuss the possibility of things being better, being like they used to be when we first met."

"Sheila, be realistic. How are things going to be the same as when we first met? We didn't have children back then, we didn't have a high monthly mortgage, we didn't have two car notes, and I certainly didn't have you trying to be my mother back then."

"Leon, I never tried to be your mother."

"Please! You're always quick to tell me what I did wrong and how it should have been done."

"But that's my way of helping you."

"Well news flash Sheila: I don't need your help on how to be a man!" He shouted.

I could have fired back at him, that is if he was such a man, then I wouldn't have to always tell him how things should be done, but I knew it would only make matters worse.

"I apologize. I didn't know I was coming off like that." I reached over and grabbed his hand. "Leon, I want us to be happy again. I'm willing to try if you are."

He didn't respond and I was concerned. I searched his face looking for answers. I found none, so I went out on a limb.

"Leon, do you want to save our marriage?"

He didn't say anything for a few moments; it was as though he was digging deep in his mind for just the right words to say.

"Sheila, you are the first woman I truly ever loved. I never knew love or was shown love before you. You changed my whole world. I tried in every way to live up to your expectations, but always seemed to fail. I just don't want to fail anymore." He said with sadness in his eyes.

I was touched and for the first time I seen the scared little boy masquerading as a man. How could I have missed that all these years?

"Leon, don't you know by now that we are as one. So if you fail, I also fail. Neither one of us can do it alone. That's why it's crucial that we work together. I need you, Leon, as my husband, my protector, my covering and my shield. I need to know that you have my back and that you will be there when all others have abandoned me, that you'll be my lover and my friend. I need to know that we'll stand together, you and me against the world. I need you to be the man I married."

Leon's expression changed with my last statement.

"The man you married! Contrary to what you may think wherever, I am, I am always the man!" He stomped out of the room, leaving me alone…again. But this time I followed him. Up the stairs we both went, until we reached our bedroom.

"You know what Leon, I've had enough of you always playing the victim. Every time I try to talk to you about us, you want to lay the blame on me. You say I am forever trying to change you and telling you how to be a man. Maybe if you were the man you were supposed to be and devoted as much time to this family as you did to your precious car or that shop, I wouldn't have anything to complain about."

"And maybe if you shut your mouth long enough and let me do what I'm going to do the way I want to do it, things would be different around here. Maybe you need to wake up and realize that the world does not revolve around you. There are other people in this universe with needs besides Sheila. You know I lied to you about where I was this past weekend. I didn't go fishing. I went to a men's retreat with Al searching for ways to be a better husband to you. But I don't think that's even possible." Leon shouted and turned to leave the room.

I jumped in front of him.

"Sheila, move out of my way or I'll move you myself!"

I've seen that anger in his eyes before so I voluntarily stepped aside. Once he was downstairs, I heard him punch a wall. He was mad beyond consolation now. Anything I said or did at this point would only make matters worse. So I stayed upstairs where it was safe. I knew I was in for a long night.

Chapter Twenty-Two
Tamara

I arrived home from work and relaxed in a soothing bubble bath, while I mentally reviewed what I wanted to say to Samuel. Afterwards, I poured myself a glass of wine and dialed his number. My heart was pounding. I was just about to hang up, when he answered.

"Hello."

"Samuel, hi. It's me, Tamara." I said trying to keep my voice from sounding nervous.

"Tamara. Well hello."

I couldn't tell from his tone if he was happy to hear from me or not.

"I missed talking to you."

"Have you?"

"Of course, but apparently you haven't missed me."

"And what makes you say that?"

"Well, I haven't heard from you."

"Nor have I heard from you."

Touché! "I guess I can't argue with that. I want you to know that I apologized to Melanie."

"You did, did you?"

"Yes, though I don't think it went over well. She wasn't in a very forgiving mood."

"Can you blame her? From where I was standing you were pretty harsh. I was shocked to hear the words you said to her myself."

"Samuel, have you even asked Melanie what happened between us that day?"

"Yes, she told me you had been calling all day and became upset when she wouldn't put you through to me. The next thing she knew you were standing in front of her going off."

"Samuel, get serious. Now does that sound like me?"

"After what I witnessed the other day, I'm not sure."

"I'm the same woman that captured your attention at the coffee shop. I'm also a woman who knows other women and I can tell Melanie has a little crush on you. That's the real reason behind all of this drama."

"That's crazy. Melanie knows our relationship is strictly professional. Besides, I don't mix business with my personal life. She in no way has anything to do with us and what we have."

Men can be so dense. If he thought Melanie has nothing to do with us, he was mistaken. She was deliberately placing herself slap dap in the middle. I had her all figured out. Samuel needed to get his head out of the sand and in a hurry!

Chapter Twenty-Three
Sheila

I felt defeated after yet another failed attempt to iron out my marital problems with Leon. I had tried it all and could think of any other recourse. I was tired of trying, tired of failing, and tired of the consequences that came with it. *His temper was becoming too much to bear. How the children slept through his ranting and raving last night is beyond me.* I definitely didn't get any rest and was paying for it this morning. I was completely exhausted—emotionally, physically and spiritually depleted. Things could not go on like this.

I feared our marriage would not be able to endure the stress and tension much longer. I needed someone to talk to before I exploded. Clyde came to mind. I'll give my pastor and friend a call and see if we could get together to discuss ways of improving my marriage. I needed some direction. Despite our one indiscretion I was confident that we could act in a mature fashion.

Though my mood was dreary, the morning was beautiful. Sunny with not a cloud in sight. *Maybe I'll suggest that we do an outdoor lunch. If nothing else, it would be safe.*

When I finally made it to the office, Tamara was already at her desk. That doesn't happen very often. I didn't recall her having an early morning meeting on today. Something else must be up.

"Good morning early bird."

"Yes, it is a great morning!" Tamara beamed.

"I know only Samuel could put a smile on your face that wide. I take it you two have mended fences."

"Yes, I guess you could say that. We talked."

"Well, I'm glad."

"Are things better between you and Leon?"

"Not really. But everything will work itself out. Thanks for asking though."

"You know I'm here if you need to talk or get drunk."

"I may just take you up on that last part."

"Anytime."

"Well, I have work to do."

"You mean e-mails to check."

"Yeah, that too," I said on my way out the door.

Just as I was heading to my desk my phone started ringing.

"Good morning. Benefits. May I help you?"

"Good morning, beautiful."

It was just the guy I wanted to hear from! The Lord was listening to my prayers.

"Clyde, what a coincidence. I was going to call you today."

"You don't really believe in coincidences, do you?"

"I suppose not. But you were on my mind."

"I hope you were thinking good thoughts."

"I was hoping we could have lunch later. I need to talk with you about some things."

"Of course, sweetheart. You know I will clear my calendar for you."

"What about 12 noon at Central City Park?"

"Central City Park? Why there?"

"I just wanted to take advantage of this beautiful day. Is that okay with you?"

"I don't foresee a problem. Like I said, I will clear my calendar and then I'm all yours."

I was a bit uneasy with Clyde's last statement, but I shrugged it off.

I was busy working when I noticed I had received an e-mail. I opened it to find an urgent message from Clyde. He wanted me to know that he was in the middle of an important project and needed me to meet him in his office. I tried to send him a response asking him to let me know when he was available for lunch, but the message was immediately returned. He apparently wasn't receiving any incoming messages. I called his extension, only to receive his voice mail. I was tempted to leave him a message to cancel, but I desperately needed to talk with him. It was already a quarter to twelve, so it was now or never.

"How about some lunch, girl?" Tamara asked.

"Not today. I've already made plans."

"Oh, okay." Tamara said a little disappointed.

"Maybe tomorrow."

"You're on." Tamara responded with a smile. She walked toward the elevators, probably on her way to the cafeteria.

I felt a little guilty about declining her lunch invitation. But I'll make it up to her tomorrow. I clicked off my computer, turned on my voice mail and went to the ladies room to freshen up.

When I arrived on Clyde's floor, his secretary was not at her desk. So I proceeded into Clyde's office. I knocked on his door. He summoned me to come in. I was grateful not to find the lights dimmed. Surprisingly, Clyde actually seemed to be genuinely working. He was pecking on his computer and appeared a bit stressed.

"If you are too busy Clyde, we could always reschedule." I told him.

"Nonsense. Just give me about five more minutes. Would you like something to drink?"

"No, thank you."

"Then make yourself comfortable and I'll be with you in a few minutes."

I sat down on the sofa and prepared to wait patiently. I turned my attention to the view outside of Clyde's window. It was breathtaking. I hadn't noticed it the last time I was in his office, but then maybe that's because he had the curtains drawn on that regretful day.

"That's a nice color on you, Sheila," Clyde commented.

"Oh, thank you." I was wearing an ice blue ruffled hem knit skirt and wrap shirt with sandals.

"Nice toes. I love to see a woman with a nice pedicure." He said not looking up from his computer.

Maybe coming to his office was a bad idea.

'You are a very beautiful woman. But you already know that, don't you?" He was now looking at me.

"Thank you for the compliment."

"I am only stating the obvious. I find you incredibly desirable. I struggle to stay away from you, but always find myself in the same predicament…and that's wanting you more."

Clyde walked over to where I was sitting. He took a seat next to me and reached for my hand.

"Sheila, this is not just a schoolboy crush. I think I've fallen in love with you. I can't stop thinking about you. I know it's wrong, but I can't help the way I feel. I want you."

He gently kissed me and I welcomed it. He pulled back, held my face in his hands, looked into my eyes and whispered that he loved me. The words sounded so good that I told him that I loved him too. He laid me back on the sofa and showered me with more kisses. The passion between us was so hot, I couldn't stop if I wanted to.

An hour later, I was back at my desk in a daze. I let it happen again. But this time I had no regrets.

Chapter Twenty-Four
Tamara

I was already making plans for my and Samuel's wedding. Never mind the fact that he hadn't asked me yet. He just had to. Because I was in love with that man. I knew I had to be truly in love with him when I started fantasizing about having his babies.

I wanted a simple but elegant ceremony, with very close friends and family. A small reception will do. That will leave the majority of the money for our honeymoon to Hawaii. My gown will be strapless and close fitting. Of course, Sheila would be my matron of honor. My ringing phone snapped me back to reality. It was Samuel. We discussed plans for dinner at my place for later on that night.

We hung up, I ran to get myself, along with the house, freshened up before he arrived. I rammed dirty dishes into the dishwasher and wiped down the countertops. I sprinkled Carpet Fresh down and quickly vacuumed the living room floor. I went into the bedroom and threw clothes that I had laying around on the floor into my closet. Then I jumped into the shower. After my shower, I put on a silk pair of red pajamas to lounge around in. I slipped into my bedroom slippers and waited nervously for Samuel to get here. *It's been so long since I have had a man in my home; I hope I don't tear his clothes off before he can get in the door good.*

Samuel arrived looking and smelling good as usual. He was wearing a pair of blue chino slacks, a yellow- striped shirt, and suede navy loafers. He definitely has great taste. He brought Chinese food and a bottle of wine. I will have to watch my wine intake, because I have a tendency to get a little loose the more I drink.

"Hello beautiful," he said as he kissed me on the cheek.

Hopefully, I will get a chance to feel his soft lips pressed against mine before the evening was over.

"Hello, yourself." I reached for the food and headed for the kitchen.

"You look comfortable. I like." He commented following me into the kitchen. "What a nice kitchen. Are these countertops granite?"

"Yes and thank you."

"Very nice." He nodded as he looked around the kitchen.

I grabbed us some plates and led him into the living room. His mouth flew open. My living room was decorated with a light velvet sofa accented with silk pillows. I had a twelve-foot long Japanese antique screen hung in the center of the wall as a focal piece with Asian accessories throughout the room.

"Wow! I see you are into the Asian motif. That picture is beautiful. Have you been to Japan?"

"Only as a child when my father was stationed there for a few years. During my visit, I really fell for Japanese décor. I've since had some pieces shipped over."

"Is your father still living?"

"Yes, he is very much alive and kicking." I said as I dished out our food into the plates.

"Does he live here in Macon?"

"No, after he retired from the military, he settled in Florida."

"Him and your mother still married?"

"No, they divorced when I was a teenager. Actually, it was during the time when we were living in Japan. Mom and I returned to South Carolina."

"I'm sorry to hear that."

"Oh, it's fine now."

"Anymore in South Carolina like you?" he asked smiling.

"No, I am an only child." I said taking a bite of my teriyaki spareribs. I was beginning to get a bit uncomfortable talking about my family. I didn't want him to know that I was estranged from my father and had very little to do with my mother. Some things are just better left unsaid and left alone. I had to find a way to change the subject.

"Would you like some more wine?" I asked him.

"Yes, please." He held up his glass as I poured. "That's good enough. I do have to drive home."

"Not really." I teased.

He looked surprised by my comment.

"I mean, I have an extra bedroom if you need it," I said quickly to save face.

"Oh, for a minute I thought you were flirting with me," he smirked.

"Remember we're taking things slow," I told him hoping he would reconsider.

"Slow is good," he said smiling mischievously.

My sentiments exactly. I cuddled closer to him and rubbed my leg against his. He stood up abruptly.

"I need to go."

"But, why so soon, Samuel? You didn't even finish your food." I placed my plate on the coffee table and stood beside him.

"I just don't want things to go too far too soon, so I had better leave." He walked towards the front door. He reached the door and turned to face me. "It's not you, Tamara, it's me. You mean too much to me for me to make a hasty mistake. I enjoyed your company." He kissed me softly on the lips and left.

Well, at least I got my wish. I felt how soft his lips are on mine, even if it was for a millisecond.

Chapter Twenty-Five
Samuel

My manhood had a mind of its own, so I knew what I needed to do. I ran. I ran as far away as I could from Tamara praying the whole time. I prayed all the way to my car, while I was driving in the car and as I was walking inside the house.

"Father, please help me to maintain power over my sexual appetite. I feel myself losing control. I don't want to fall…again. Deliver me from the temptation of wanting to please my flesh. Strengthen me, Lord, for I am weak."

It was embarrassing to leave from Tamara's place so abruptly, but I had no other choice. I felt myself slipping. Tamara has such an effect on me. I was really falling for this woman. But I was sincere when I told her that I wanted to take things slow. I didn't want history to repeat itself. What happened between Rachel and me still haunts me, even though it's been over three years ago.

Rachel was the love of my life. We met at a book signing in a local bookstore. We both loved Nikki Giovanni and she was in town promoting her latest book of poetry. I noticed her eyes from across the room. They were a beautiful shade of light brown and so full of life, much like Tamara's. I truly understood the saying "The eyes are the windows of the soul," when I looked at her. Her eyes spoke multitudes and made me want to know her more.

She was talking with another woman and hadn't noticed me staring at her. I had to get closer to her. I wouldn't allow myself to leave without knowing her name. So I stood back and watched her every interaction waiting for an opportunity to introduce myself. It was apparent that she frequented this particular bookstore often. She appeared to know just about everyone, or maybe she was just that friendly. She definitely had a presence about her. I started to make my way towards her. I was standing directly behind her and was about to tap her on the shoulders when she suddenly turned to face me. I was frozen like a deer caught in headlights. She smiled and said "Hello". Just the sound of her angelic voice melted my heart. I was in love.

"Hi", I responded.

"Don't you just love book signings?" she asked with

excitement. "I'm Rachel Davis by the way."

"I'm Samuel. Samuel Roberts," I answered clumsily.

"Don't tell me this is your first time?" she smiled.

"First time?"

"Book signing", she pointed to the book in her hand looking puzzled.

"Oh, no. But it is my first time at this location."

"Well let me take this time to officially welcome you, Samuel."

I was captivated by her. It took me a moment to notice that she had her hand held out for me to shake.

"I'm so sorry. Thank you," I said as I reached for her hand.

She had long, slender fingers and her skin was very soft. I didn't want to let go.

"I tell you what, Samuel, seeing that it's your first time visiting us why don't I arrange a one-on-one with you and Nikki after the book signing?'

"You can do that?"

"Of course, I'm the store owner and a close personal friend of Nikki's."

"I would love it! She is my all time favorite!"

"Then it's done. Just meet us in the café section of the bookstore in about an hour."

That was the beginning of a sweet love affair, our first date and one of many. Rachel was the wind beneath my wings. She supported me when I quit my job as Financial Manager with Geico to start my own firm and helped me through the process. She had plenty of contacts and actively dropped my name to a few letting them know there was a new accountant in town and his name was Samuel Roberts. She changed my life in more ways then she could ever imagine. I wanted to make her my wife and partner for life.

But like most relationships, no matter how wonderful, tragedy is bound to happen. During our courtship Rachel became pregnant. I was ecstatic to hear this news and was prepared to marry her. But Rachel had other plans and they didn't include marriage or a child. Without consulting me, she decided to have an abortion. I pleaded with her not to take this drastic step. I even said that I would raise the baby on my own. I tried everything, but in the end against my wishes she terminated the pregnancy. Soon afterwards, she closed the bookstore and left town. I never saw or spoke with her again.

The loss of both my unborn child and Rachel left me devastated. I felt as though my heart was hemorrhaging. I had hit rock bottom and didn't want to go on. My business was failing, some mornings I didn't even bother to go in. I didn't have the strength to face the day. My Rachel was gone. I hadn't slept in days and was at the end of my rope. I was tired of the pain and wanted relief.

One night I was alone in my apartment with nothing, but a bottle of prescription sleeping pills. I was equipped with all that I needed to permanently ease my suffering. But before I could take an action something made me pause. It was coming from the television. I stopped to listen.

It was an evangelist and he said that God loved me. He also said that God had plans for my life. I listened to the remainder of the broadcast and as a result, that night I did die as planned. But it was the old me. I opened my heart and received Christ that night and it was the foundation of my new existence. I invited God into my situation and vowed never to do anything without His guidance.

By no means did I want to go through what I went through with Rachel ever again. The only way I could ensure that was to do things the correct way from the beginning. No sex before marriage, just as God had intended. Though some days were harder than others, but I was committed to my decision. Now if only I could get Tamara there with me.

Chapter Twenty-Six
Sheila

It was the end of the day and I was about to clock out when my phone rang. Usually with it being so close to quitting time I would not have answered. But I was in a good mood, so I picked up.

"Benefits. How may I help you?"

"Hey baby."

"Mama Lyn!"

"How's everything sugar. Mama Lyn hasn't heard from since that weekend I kept the children for you."

"Everything's the same, Mama Lyn", I sighed.

"You mean Leon's still hasn't come home?"

"No, he came home. I did what you told me to do but we still ended up arguing. I'm tired Mama Lyn and I just don't care anymore. I have decided that he can continue to do whatever he wants and so will I."

"The devil is a lie! You can't leave your marriage to chance and you can't expect things to get better overnight. You have to pray and be still. It's when you are quiet that you can hear what the Lord is saying to you. He hasn't forgotten about you honey. He has heard your every prayer. Your job now is to be patient and wait."

I was on the verge of tears and I didn't want to cry at work. Tamara would have a million questions and I didn't want to have to explain.

"Mama Lyn, I can't talk about this at work."

"Okay baby. I understand. But let me leave you with one more bit of advice. God created marriage. One Adam for one Eve. Don't let anyone come in between what He has put together. Don't let your flesh get in the way of your breakthrough. It's not worth it. Mama Lyn loves you baby."

"I love you too." A tear trickled down my cheek.

"I'm still praying for the both of you."

"Thank you, Mama Lyn."

I don't know how she does it, but she always has a way of knowing what's going on with me without me telling her. It's as though she's some type of psychic or something. She has been that way for as long as I can remember.

My mother passed away of breast cancer when I was seven years old. Mama Lyn took my younger brother and me in and raised us. I don't remember much about my mother besides her being sick most of the time. My father wasn't around much. He sought refuge through other women and alcohol. So Mama Lyn was all we had. She took care of mommy until her demise. It was only natural that she would take care of us too.

Mama Lyn was a deeply spiritual woman and believed in praising God. Every morning we would wake up to gospel music blaring from the radio and Mama Lyn singing along. She taught us to pray and to thank God for everything, even the rain. She said without the rain the flowers wouldn't grow. As in life, without trials your faith wouldn't either. Mama Lyn was full of faith. So it was good to know that she was praying on our behalf. Because if anybody could get a prayer through it was certainly Mama Lyn.

On my ride home Mama Lyn's words echoed through my mind…*Don't let anyone come in between what God put together.* Anyone. That included Clyde. Guilt washed over me and I began to cry. I took things into my own hands once again instead of waiting on God. I had to be patient. But first I needed to repent.

"Lord, I'm sorry. Please forgive me. I know what I did was wrong. I promise to never run in front of you and try to handle my problems on my own. From this day forth I'm going to wait on you! Amen."

I made it home, greeted my children and my husband, then I asked Leon if we could pray. He was shocked at my request. Actually he laughed. But I was determined that the devil was not going to win this fight. He couldn't have my family. I'm taking back what was rightfully mine. Leon belonged to me. God gave him to me and wasn't no devil in hell going to steal him away. I was prepared for battle!

Chapter Twenty-Seven
Tamara

I was exhausted and almost missed the beauty of the sun hanging perfectly in the sky greeting the day with her presence. My mind was too preoccupied with thoughts of Samuel to fully appreciate it. The sun was ushering in the start of a startling new morning and I barely noticed.

I thought surely Samuel would have called me when he arrived home from basically sprinting out of my front door the night before. But I hadn't heard from him. *What's wrong now?* He and I couldn't seem to get past first base with this relationship. Just when I thought we were making progress a curve ball was thrown. I wondered if I came on too strong the other night.

The more I thought about him, the more I needed to see him. I pulled out my cell phone and called Sheila. I told her that I had a few errands to run and would be a little late coming in. My first stop being Roberts' Accounting Service. I wanted to talk with Samuel face to face.

When I arrived Melanie wasn't at her post. I went directly to Samuel's office. When I appeared at his door I found that he wasn't alone. I observed him embracing a woman. As I got closer I realized the woman was Melanie! My instinct was right! The little witch was trying to steal my man. I knocked firmly on the door to get their attention.

"I hope I'm not disturbing anything," I sneered. If looks could kill, they both would have been laid out on the floor. At the sound of my voice they broke their embrace. That's when I noticed that Melanie was crying. She turned her back to me.

"Tamara, what are you doing here?" Samuel asked me as he handed Melanie a tissue from off his desk.

"I'm here to see you. But I guess I should have called first." I answered sharply.

Melanie never uttered a word as she brushed past me. I was getting pissed by the minute and wanted some answers. I was tired of men trying to make a fool out of me. I wasn't leaving until I got to the bottom of this. Samuel needed to make a decision: it was going to be either her or me.

"So is she the reason why you left in such a hurry last night?" I asked not realizing I was raising my voice.

"Tamara, it's not what you think," he started to explain as he closed his office door. "Please don't jump to the wrong conclusion."

"Well what am I supposed to think Samuel? Melanie doesn't like me for no good reason, you bolt out of my house before we can even eat dinner last night and now I find the two of you in each other arms today!"

"Tamara, please have a seat," he gestured for me to sit down. I reluctantly complied. I folded my arms across my chest and waited for him to speak. He appeared nervous.

"Just say what you have to say Samuel. I'm a big girl and I can take it." I said with more boldness than I really felt.

He stared me in the eyes. I felt a sense of seriousness. *I hope he's not going to break things off between us. I don't think I would be able to take it!*

"I have something very important to tell you," he finally said.

My mind was spinning, wondering what in the world he was going to say. *Please don't let him be gay!*

"There's no other way to say this, but to just spit it out. I'm celibate!"

Celibate! Did I hear him right? Now I knew why Melanie was crying. I felt like crying too.

Samuel went on to tell me about the relationship he had with some chick named Rachel and the outcome of it. He confided in me about his suicide attempt and how he accepted the Lord into his heart that night. It was then he made a vow to never have sex before marriage again. I sat there and listened to every word. When he finished I didn't know what to say. I didn't know whether to hug him or to run out the door. I've never known a man who had given up sex, especially one that I was dating. This was definitely a first. What would that mean for us…for me? *Does this mean I have to give up sex too?*

"But wait a minute. What does this have to do with me finding Melanie crying in your office and you consoling her?"

"That was something totally separate from what I just

shared with you. Melanie approached me about a relationship between the two of us. I told her that we could never be. One, because we worked together, and two, because I am in love with you."

For a few moments I believed I actually stopped breathing. *The man of my dreams just professed his love to me, but he was also celibate and off limits to explore that love. Who said all was fair in love?*

Chapter Twenty-Eight
Sheila

Even though Leon had agreed to pray with me, his heart didn't seem to be in it. He wouldn't lead any of the prayers, but insisted that I did. I'm not sure why Leon was holding back from me, but I hope it didn't involve another woman.

I realize that I had broken my marriage vows with Clyde, but I still didn't like the thought of Leon being with someone else besides me. Someone else hearing his dreams, someone else soothing his fears, someone else having his heart was unthinkable. That was my job. If another woman was involved, I prayed that the Lord would remove her just as I asked Him to remove Clyde from my life.

Clyde had become a dangerous distraction for me. He was a drug I was getting addicted to, and like all drugs he was hazardous to my health. I had begun to rely on Clyde to fix the problems in my marriage and to make me feel better when things weren't going right. I had made him my God.

Tamara had called to tell me that she was coming in late, so I went down to the cafeteria to grab some breakfast. I switched on my voice mail and headed for the elevator. As I was waiting for the elevator, Cindy Harper from Payroll walked up.

"Hey Sheila," Cindy said showing a perfect set of teeth. Cindy was originally from Arkansas, in her early forties, had been married and divorced twice and was currently on husband number three. She was attractive, but flamboyant and loud. She also knew all the gossip about every employee in the company and didn't mind sharing it.

"How's it going over in Payroll?"

"Oh, it's going."

The elevator doors opened and we entered. To my surprise Clyde was inside. He looked just as surprised to see me.

"Good morning ladies," he smiled.

"Good morning," we both answered in unison.

"Which floor would you ladies like?"

"I'm going to the first floor," I answered.

"Me, too," Cindy added. She was smiling so hard, I

thought her face would crack.

Clyde pressed the first floor button for us.

"You ladies are looking quite lovely today," Clyde commented.

"Well, thank you Clyde. You're not looking too bad yourself," Cindy said enticingly.

The elevator doors opened. Thank God. I didn't think I would be able to take much more of Cindy's flirting.

"This is my floor. You ladies have a wonderful day," Clyde said exiting the elevator.

"I will now," Cindy hollered out the door. She was so embarrassing.

"That Clyde is one sexy man. I would probably give him some play if he hadn't been with just about every woman here."

Her comment hit a nerve with me. I was sick of her and all her rumors.

"You don't know that, Cindy."

"Oh, yes I do and I also know that he's messing around with Julia right now!" she boasted.

Now, she had my full attention.

"Julia Burns who works in Purchasing?"

"That's the one."

The elevator doors opened to the first floor. We both walked toward the cafeteria, but I had suddenly lost my appetite.

"How do you know for sure?"

"Because Julia and I work out together and she told me." She was smiling from ear to ear as though telling me this information was bringing her pleasure.

Cindy kept rambling on about this and about that but I didn't hear her, because my mind was on Clyde and Julia. So he had seduced her too! *And to think I was foolish enough to believe that he actually cared about me. He told me that he loved me. Had it all been a lie? I needed to know the truth and I needed to know now.* I supposed I should take running into Cindy and her so graciously revealing Clyde and Julia's affair as confirmation from the Lord. But it felt more like a gut punch.

"Hey girl, I don't have the stomach for cheese grits this morning so I will see you another time."

"But they're serving omelets and you don't want to miss those," Cindy bellowed.

"Maybe next time." Her big, loud mouth was really starting to get on my nerves.

I left her standing in the omelet line and headed for Clyde's office. I made it in record time. I walked right past his secretary and directly into his office. He was on the phone, but hung up as I closed the door.

"Sheila, what a wonderful surprise," he said as he stood up to greet me. He walked toward me and I began to get weak. Just the scent of his cologne made me dizzy. He grabbed me by my waist and pulled me to him.

"Are you seeing Julia?" I blurted out before I lost my nerve.

"What?" He took a step back.

"Julia Burns. Are you seeing her, or shall I say having sex with her?"

"Let me guess Ms. Mouth of the South told you that tidbit of news this morning after our elevator ride together."

"Is it true?"

"Sheila, I am aware of all the innuendoes and rumors people say about me. But please don't make me explain every one of them to you. Don't put me through that."

"I won't put you through anything because I'm finished." I removed his hands from around my waist.

"Sheila, don't do this. I need you."

"Need me for what?"

"To fill my days."

"That's what your wife is for or have you forgotten." I turned to walk out the door, but he grabbed me by the arm.

"Don't leave me, Sheila. I love you."

The words sounded so sweet. He stood behind me and kissed the back of my neck. He was reeling me in.

"Don't I make you feel good?" he asked in between kisses.

I was falling into his trap again. I turned to face him and suddenly his phone rang. Coincidence or not, nevertheless, the interruption snapped me back to my senses.

"Clyde, this is over. I love my husband and I want my

marriage to work. After today don't call me or come to my office anymore." I spun on my heels and left his office before he had a chance to respond.

I prayed that with each step I took away from Clyde, that it was a step closer to Leon. I knew it was going to be hard to break loose from this emotional tie Clyde had over me. But with prayer I knew the Lord would see me through. He had too! Because I was helpless on my own.

Chapter Twenty-Nine
Tamara

I couldn't wait to get into the office to let Sheila in on my new discovery regarding Samuel. I still couldn't quite wrap my own head around it. This was something totally new for me. I've had men only want me for sex. I have even used a few for my own sexual gratification. But I have never run across a man who wasn't having sex at all!

What timing! Sheila was entering the elevator just as I was.

"Good morning, sunshine!"

"Morning."

"I have some juice for you." I said excitedly

"On who?" Sheila asked wide-eyed.

"Samuel!"

"Don't tell me he asked you to marry him?"

"I wish!"

The elevator doors opened to our floor.

"Come to my office so we can talk," I told her.

"Let me check the voice messages and my e-mails, then I'll be right in."

"Well, hurry up! I'm about to burst."

"In that case the voice messages can wait." Sheila followed me into my office.

"Door opened or closed?" Sheila asked.

"Definitely closed."

"Girl, spill the beans." Sheila said as she sat down in her favorite chair.

"Samuel let me in on a little secret of his and I'm not quite sure how to handle it." I said as I sat in the seat next to her.

"What? Is he bisexual?"

"Girl, its worse than that." I lowered my head.

"Worse than that? Don't tell me he's one of those down low brothers."

"No." I paused. "He's celibate."

Sheila just stared at me.

"Did you hear what I just said?"

"Yeah, I'm still waiting for the bad part."

"Sheila that is the bad part! He's practicing celibacy and I'm not. What could be worst than that?"

"Honey, there are plenty things worse than your man being celibate. Try a drug dealer, a child molester or a cross dresser just to name a few. Would you prefer those options? Shucks, I can handle celibate."

"Maybe for you. But I enjoy having sex. It's my reason for living."

"Your reason for living? You sound like a teenager who just discovered an orgasm. Sex is good only if you care about the person that you're with. What is so difficult about getting to know Samuel for who he is and letting him know you on a deeper level?"

"And what's wrong with two people just getting it on like grown folks do? He must think there is something wrong with me."

"Who says something has to be wrong with you?"

"Then, why doesn't he want me like I want him?"

"Tamara, him wanting to wait to have sex tells me he cares a great deal about you. Any two people can have sex together. But Samuel doesn't want just the physical aspect. He wants all of you—the whole package."

"But I don't know if I can wait. Just the sound of his voice makes me hot. Sheila, you know me. I believe people should do what they feel and worry about the consequences in the morning. So a man not interested in taking me to bed is new for me. I haven't even talked to him since he divulged this bit of information to me. I don't know what to say. Sheila, what am I going to do? I don't want to lose him."

"Then, take it to the Lord in prayer." Sheila said.

"Prayer?"

"Yes, prayer."

"And what am I going to pray for? That Samuel comes to his senses and makes love to me? I don't think God would answer that prayer."

"You can ask the Lord to help you with your self control."

"Sheila, please. Why would God care about my self control?"

"Everything about living for God is about self control. I

have to ask Him to help me each and every day."

"You? Besides Mama Lyn, you are the most righteous woman I know. Why would you need help with self control? Sheila, you already have it all together. I only wish I could be more like you."

"Well, believe it or not I do have struggles."

"What struggles are you talking about? You are a wonderful wife and mother. From where I'm sitting your life is perfect."

"No, my life only appears to be perfect." Sheila said with sadness in her eyes.

"Sheila, what is it?" I sat on the edge of my chair and grabbed her hand.

"Tamara, my life is not perfect and neither am I. I made a very big mistake. A mistake that could cost me my marriage if Leon found out."

"What mistake?"

"I'm so ashamed."

"Don't be. You know you can tell me anything and I will never judge you."

I couldn't imagine what she could have done so wrong, but in the back of my mind I knew it had something to do with Clyde.

"Sheila, please talk to me."

"I…I slept with Clyde."

What I feared most had come true. Sheila had let the serpent into her garden. The woman I looked up to had slipped and fallen.

"Please don't be mad at me, Tamara."

I quickly embraced Sheila. The look of desperation and shame on her face overtook me. I couldn't be mad at her. I felt sorry for her. Clyde used her weakness to slither his way in.

"How could I be mad at you? You just made a mistake. You're human."

"I feel like such a failure and a hypocrite. I knew better," Sheila said.

"Sheila, you went to Clyde for help and things got out of control. If anyone is a hypocrite, it's him. He's the pastor!" I was beginning to get angry.

"It's not entirely his fault. I was a willing participant. I could have stopped him, but I didn't want to."

"I didn't have any idea that things were that bad between you and Leon?"

"Things have been. But now I'm determined to seek God regarding my martial problems, instead of working on them with my own strength. I ended things with Clyde and devoted myself to praying with Leon on a daily basis. Our situation is bound to get better with God in the middle. I hope what I told you doesn't make you disappointed in me."

"No, of course not, I'm just shocked. You stopped letting me in."

"I know and I'm sorry for that. But I felt you wouldn't understand and Clyde was always so willing to listen to me. He was so attentive and encouraging in the beginning. But then he started getting a little flirty and I was flattered. Leon wasn't paying any attention to me anymore, so I welcomed his advances."

"Clyde knew exactly what he was doing all along and you played right into his hands. You may have been a willing participant, but you were also a victim."

"I thought he really cared for me, Tamara. I feel so violated."

"I know exactly how you feel." Sheila looked at me with questions in her eyes. "Sheila, I never told you but when I was a little girl one of my uncles molested me. He befriended me just as Clyde did with you. He caught me when I was in a low state in my life. The kids at school picked on me, my mother was working long hours and was never home. But I could always count on my Uncle Ray to be there. Before long, one friendly gesture turned into another one. He seduced me with his conversation and then when my guard was down he raped me. It all seemed to happen so fast, I didn't see it coming until it was too late. He took something I can never get back, my trust. I don't think I will ever forgive him for that."

"Tamara, I am so sorry."

"Sheila, Samuel is the best thing that has ever happened to me. Before him, I never thought I was capable of really loving a man. All my life they only brought me hurt and disappointment, but not Samuel. He entered my world and gave me a new perspective on life. I don't want to lose him."

"Then, you need to pray."

"God isn't going to hear anything I have to say, Sheila. I doubt that He even loves me."

"Nonsense. He does love you. He loves you so much He allowed His only child to die in your place. Do you believe that?"

"Whatever you say."

"You have to believe in your heart that Jesus died and God raised him from the dead, so that you and I could live!" Sheila grabbed me by both hands and looked me eye to eye. I felt the intensity of her words. "When you believe you have already won the battle and the devil has no power. We have to pray daily for God's guidance and direction to overcome the devil's schemes he tries to plot against us."

"Are you sure He can help me?"

"Girl, He is the only one who can help us both."

I wasn't absolutely sold on the idea that God could or would help me. But I did know that I needed all the help I could get. Because things aren't right in my life and hasn't been for awhile now.

Chapter Thirty
Sheila

My week had turned out to be a good one. Leon and I were still praying together at home. He even led our last session and it was such a heartfelt prayer that I was moved to tears. Tamara was making progress with Samuel and their relationship appeared to be going strong. Yes, the Lord was indeed hearing and answering our prayers. Life was good.

The weekend was approaching and I had invited Tamara over to our house for dinner. It was high time that we met this Samuel in person. But from what I learned about him from Tamara, I liked him already. I only hoped him and Leon hit it off. Leon wasn't the dinner party type. He was more into house parties with finger foods, alcohol galore, and loud music. That's why I was surprised when he agreed to have Tamara and Samuel over. My honey was definitely trying to please me.

Tamara returned back into in the office with a big grin plastered on her face. I assumed she had met Samuel somewhere for lunch. I was so happy that she had finally found happiness. She surely has had her share of bad relationships. It was turning her cynical about all men. But after kissing so many frogs, she now had her Prince Charming.

"I take it we had a nice lunch," I teased Tamara as she passed my desk.

"Yes, we did."

"I'm glad. What are you guys going to have a taste for this Saturday night?"

"Oh, it doesn't matter. You know I enjoy anything you prepare."

"Well, what about Samuel. Is there anything that he doesn't eat?"

"I don't think so. Besides that man is from the dirty south. You know we eat anything."

"Ummhmm…so if I cook up some chitin's he'll eat that?"

"I don't know about Samuel. But as long as you have some hot sauce on hand, I'll eat em!"

I laughed at Tamara's comment. As high fashioned as she

appeared to be that girl was country to her heart.

"So what are you cooking?" Tamara asked.

"I'll just surprise you."

"You sure you don't want me to bring anything."

"Nope. I have it covered."

"Well, what if I just bring a dessert."

"Store bought?"

"Of course."

"Don't bother. I'll throw something together."

"Suit yourself. But ain't nothing wrong with Mrs. Smith. You haven't lived until you tasted one of her deep dish peach pies."

"I think I can bake up one better."

"Well go on with your bad self. I don't doubt your skills. Remember I have had your cooking before. My mouth has been watering since the day you invited Samuel and me over. I can't wait to sit at your table. Leon better be careful or I might move in."

"Girl please. There is no way I can feed your hungry self on a regular basis. I don't understand how you keep your weight down."

"It's called exercising four times a week."

"Oh, I wouldn't know. I guess you single women have time for that, but I don't."

"You have to make the time."

"And would that be after I cook for my hungry crew, before I help the children with their homework or after I iron everybody's clothes. There are just not enough hours in the day."

"You need to give Leon some of those chores. It's his family too."

"One thing at a time, Tamara. He's coming around."

* * *

It was already 2:00 and I was just leaving out of the grocery store. I couldn't choose between the leg of lamb or red snapper. I decided on the red snapper. I would prepare my mushroom stuffing and spinach casserole, along with a peach cobbler to die for. I was going to prove to Tamara that Mrs. Smith don't have nothin' on me. I couldn't go wrong with Mama Lyn's recipe. She probably taught Mrs. Smith a thing or two!

I hurried home to get dinner started. I had told Tamara to be at the house by 6:00. Leon got home from work a little after 3:00. He was dirty and grimy from fixing on cars. Leon had a reputation for being one of the best mechanics around town and Saturday was one of his busiest days at the shop. But he promised he would be home early in time for dinner. I had my red snapper and mushroom stuffing in the oven and was starting on the spinach casserole.

"Mmmm…that smells good, baby," Leon said as he kissed me on the cheek.

"Thank you honey. How was your day?"

"Busy. I'm going upstairs and soak in the tub before dinner."

"Call me if you need any help washing anything."

"Don't tease me girl," he smacked me on the butt as he walked upstairs.

"Who said I was teasing," I yelled after him.

Thank goodness we are once again the happy couple we used to be. It's been a long time coming. But Mama Lyn was right. I had to give it to the Lord completely and not take it back. I had been good at praying and then as soon as things didn't change as fast as I thought they should, I would try to handle it on my own. I didn't realize I was standing in my own way and blocking my blessings. But those days were gone. I was now standing on the promises of the Lord.

I was mixing my peach cobbler filling together when I realized I was out of cinnamon. A peach cobbler isn't a peach cobbler without cinnamon. I heard Leon let the water out of the tub some minutes ago. I was going to have to send him to the store.

"Sweetie I am all out of cinnamon. Could you drop by the store and pick up some? You can take the kids by Mama Lyn's while you're out."

"Yeah, no problem. Are they ready to go now?"

"Yes, they should be."

"Okay, then I'll be back in a jiffy." He kissed me before leaving.

Everything was done, except my peach cobbler and Leon still wasn't back yet. I decided to go ahead with my shower. It was now after 5:00 and still no sign of Leon. I was dressed and downstairs waiting on Leon, Tamara or whomever showed up first. The doorbell rang causing me to jump. It had to be Tamara and Samuel, because there was no reason for Leon to be ringing the bell. I put on my biggest smile and went to answer the door, but before I could, to my surprise in walked Leon, along with Tamara and Samuel.

"You look good enough to eat sweetheart," Leon said locking his lips on top of mine. "I stopped and bought us a bottle of wine. I had to go to three different liquor stores to find that special brand that you like. Here's your cinnamon, compliments of Mama Lyn," he handed me the cinnamon. I smiled to myself thinking how good it felt that Leon had gone out of his way for me…just because. God was answering my prayers.

Chapter Thirty-One
Tamara

"Wow! What a scrumptious dinner. I do believe if Sheila wasn't already taken I would have to ask her to marry me!"

"Watch out now! It was good, but it wasn't that good!"

"Woman please! You were there! You tasted it! Sheila put her foot in that meal. Can you cook?"

"What if I can't?"

"Then maybe you should ask your girl for some lessons," Samuel joked. I looked at him and rolled my eyes. "I was just kidding baby. Not about dinner though, it was slammin'."

"So what did you think of Leon?"

"He seemed like a real jokester."

"Yeah, Leon is always the life of the party."

"He was trying so hard to be comical that I couldn't get a real feel of who he was. But I can see why you are so crazy about Sheila. She seems to be a very nice lady. I can tell she has a good heart."

"That she does. She's always taking care of everybody else's needs first and she loves her family."

"It shows."

"So you had a good time?"

"Tamara, anytime I'm with you I have a good time."

I smiled to myself. *Man, I wish he would come home with me!*

"What are you smiling about over there?" Samuel asked placing his hand on my knee.

"Trust me, you don't want to know."

"This is hard for you, isn't it?"

"What?"

"You know what, the two of us abstaining?"

"Honestly, yes. But that's because I've never had to do it before or even thought about it for that matter. I'm accustomed to expressing how I feel to the person I am seeing."

"Well, there are other ways of expressing to someone how you feel besides sex."

"Really now?" I asked him in surprise.

"Yes, really."

"Pray tell."

Samuel looked at me with a wicked grin and continued to drive. He drove right past my neighborhood and I started to get excited. *Is he taking me to his place?*

"Where are we going?" I asked him.

"You'll see."

We drove on for what seemed like an eternity. He finally pulled into a wooded area. He stopped the car, got out, walked to the back of the car and went inside the trunk. He pulled out something that looked like a box. He came over to my door and opened it.

"Right this way my fair lady." He held his arm out for me to grab and with the other hand he was carrying what looked to be a picnic basket.

"Samuel?"

"No questions. Just follow me."

I followed him through a pathway that lead into an open field. When we reached the appropriate spot he spread out a blanket which he took from out of the basket he was carrying.

"Have a seat please."

As I sat down he pulled out wine, wineglasses and fruit from the basket. He poured us some wine.

"I hope you're thirsty," Samuel's said as he handed me a glass of wine.

"Samuel," He placed his finger over my mouth.

"Don't say anything. Words are not needed. I brought you here to enjoy the scenery," He then fed me a grape.

I took a sip of the wine and looked up at the sky. It was a spectacular sight. Every star was out tonight and it seemed as if they were dancing just for us. It felt magical. I sensed Samuel felt it too, because he snuggled closer to me.

"Tamara, I hope you know just how much I love you. You are everything that I have been wanting and waiting for all my life." Samuel whispered in my ear.

I was growing with excitement with each word he spoke! He was finally ready to take our relationship to the next level!

"This little corner of the world is a very special one to me and I wanted to share it with you, because you have grown to be very special to me too. Tamara, I have wanted to ask you something very personal for the longest time, and I hope you don't say 'No'."

"Samuel you can ask me anything," I panted.

"I wanted to know if you would attend church services with me on Sunday. It would mean so much to me."

CHURCH! I can't believe all this was for him to muster up the courage to ask me to go to church with him. Here I thought we were going to make passionate love under the stars!

"Well, of course, Samuel. I would love to attend church with you," I answered pasting on a fake smile.

"Great! I'll pick you up in the morning around 10:30," he said smiling from ear to ear.

At least one of us is getting some satisfaction tonight.

Chapter Thirty-Two
Sheila

I was in my car on my way to work when the words to the song, 'Oh what a beautiful morning' came to mind. It had been a perfect weekend. I couldn't have planned it better even if I tried. The dinner Saturday went off like a charm. I loved Samuel. He was every woman's dream. Tamara's done good! Sunday after church Leon took me out to dinner and a movie before we picked the kids up from Mama Lyn's. He's been so wonderful lately and I was falling in love with him all over again. I just couldn't get enough of that man. As I was waiting for the elevator Monique walked up. I had not seen her since that day in Tamara's office.

"Hey girl," I greeted her.

"Good morning," she responded.

"A good morning it is."

"I take it that things have gotten better for you," Monique asked.

"Much better," I smiled.

"You thought it through?" She asked.

"I prayed it through!" I answered.

The elevator doors opened to my floor and I exited. Just as I was checking the weekends voice mails, Tamara entered the office. "I brought doughnuts."

"You're too good to me," I joked.

"And don't you forget it woman," Tamara laughed.

I grabbed a doughnut. "No milk?"

"There's just no pleasing you. I don't know why I try," Tamara teased.

"Because you love me."

"Lucky for you. So how did you like Samuel?"

"He seems to be the real deal, Tamara. I like him a lot."

"Good, because he loves you!"

"Really?"

"He went on and on about how good dinner was. Almost got on my nerves!"

"Tell him he is more than welcome to come to dinner anytime."

"Oh no, I am breaking this up now! The next time we are all going out for dinner."

"You're a mess girl."

"You will never guess where I ended up yesterday," Tamara said.

"Not Samuel's place!"

"Heck no. I went to church with him."

"You did. How did that happen?"

"Girl, after we left your place he took me to this secluded area and we had a picnic beneath the stars. It was so romantic. It had to be the most thoughtful thing any guy has every done for me. He told me how much he loved me and never wanted to be without me. Then he asked me if I would go to church with him."

"You sound disappointed."

"I just have to get used to all of this. Remember this whole no-sex thang is new to me. When we arrived at the field and he had laid everything out and was whispering in my ear, I just knew he was going to lay me out next!"

"Girl, stop! I thought you always wanted a guy who was more interested in knowing you personally than sexually."

"I did. I mean I do. It's just I never thought such a guy existed."

"Well, he does and his name is Samuel. So what are you going to do about it?"

"Continue to take cold showers and eat plenty of chocolate. And ironically guess what the sermon was about on Sunday?"

"What?"

"'Is Your Bed Honorable', taken from Hebrews 13:4."

"Hebrews 13:4. What's Hebrews 13:4?" I wasn't as well versed to the books of the bible as Mama Lyn.

"I quote, '*The marriage is honorable in all and the bed undefiled.*'"

"Was that a sign or what?" I said to Tamara unable to hold back my laughter.

"God is definitely trying to tell me something," she said as she walked towards her office.

After Tamara left, I took my Bible out of my desk and turned to Hebrews 13:4. A cold chill ran through me as I read the rest of the scripture…*but fornicators and adulterers God will judge.*

Whoa! Okay, I hear you God. Loud and clear.
My phone rang. I closed my Bible and answered it.
"Good morning, Benefits. May I help you?"
"Good morning, my sweet."
It was Clyde.

Chapter Thirty-Three
Tamara

I made it to my office and turned on my computer to check my schedule for today. I was silently praying that I didn't have a day jam packed with meetings. I received a signal notifying me that I had mail. I clicked onto my e-mail. It was an e-greeting card from my honey. He was thanking me for a wonderful weekend! *Samuel was so thoughtful. I don't know what I have done to deserve such a magnificent man, but I am thankful.* I was about to send him a reply when my phone rung. I started not to answer it, but I was at work.

"Good morning, this is Tamara Washington. May I help you?"

"Good morning, gorgeous."

"Hi handsome."

"How is your morning going so far?"

"I can't complain. As a matter of fact I was just about to reply to the e-card you sent me."

"No need for that. You can tell me right now what you wanted to say."

"Just thank you for being so thoughtful and considerate. You have been nothing but a wonderful gift to me Samuel and I appreciate you."

"Being good to you comes easy because I love you so much, Tamara."

"I love you too."

"I felt like a king with you by my side yesterday at church. Did you see all the other brothas staring at you?"

"No, I was too busy watching the sistas."

"They were all probably wondering who that beautiful, mocha queen was with Brother Samuel, especially with you being the first woman I ever brought to church."

"Really?" I felt special knowing that I was the only woman Samuel had escorted to church realizing how important the church was to him.

"You are going to understand one of these days just how truly special you really are to me. So do you have any lunch plans today?"

"Not yet. I was just about to check my schedule for today. Why? Did you have some place in mind?"

"Actually I am having a working lunch meeting with a possible new client. I was just wondering what you were doing."

"I will most likely do lunch with Sheila. By the way she just loves you and told me to tell you that you are invited over for dinner anytime."

"Really now!"

"But I told her we were declining the invitation, because the next time we all were going out to a restaurant for dinner."

"Yes ma'am!" Samuel said laughing.

"I'm serious."

"I know you are that's why it's so funny. But you know I love you despite the fact that you can't cook."

"I can cook!"

"Sure you can sweetheart. I hope the rest of your day is great." Samuel said still snickering.

"Yours too, and I wish you well at landing your new client."

"Thanks honey, see that's why I love you so much."

We hung up and I went to ask Sheila if she had any plans for lunch.

"Hey you. Do you have any lunch plans today?"

"Not really," Sheila answered with an odd look on her face.

"Are you okay? You look as though you have seen a ghost."

"More like heard from a ghost," she responded.

"What do you mean?"

"Clyde. He called me this morning."

"He what? Well I know what he wanted, but what did he say?"

"I didn't give him a chance to say anything. I hung up before he could say a word."

"Sheila, it seems as though he is not going to go away quietly. You may have to tell Leon."

"Have you lost your mind? I can never tell Leon!"

"But Sheila, all I'm saying is…"

"Forget it Tamara! I'm not telling Leon. I will just have to handle this on my own! Conversation over!"

I had never seen that look in Sheila's eyes before. It was a look of hopelessness. I left her alone, because I knew anything I said would fall on deaf ears. I walked back into my office and began to work. Hours later Sheila burst through my door. It scared me to death. Because I wasn't sure if she was going to light into me like before or if she was going to reveal to me that she had done something crazy to Clyde.

"Your lunch plans are solved!" Sheila exclaimed!

She walked into my office with two delivery men following behind her who were both carrying boxes.

"What's going on?"

Sheila handed me a card as the two guys started to unpack the contents of the boxes onto my desk. The card read:

Sorry I couldn't join you for lunch today.
But know that I am with you in spirit.
Enjoy.
Love Always, Samuel.

Samuel had lunch sent in to me and from my favorite restaurant, Taylor Made Grill! *Oh, how I love this man!* The men had finished unpacking lunch and I noticed the setting was for two.

"Sheila, it's two of everything! I guess your lunch plans have been solved too. Bon appetit!"

"Tamara, that Samuel is one of a kind. God has really shined His light on you."

"I'm speechless."

"And with good reason. All the years that Leon and I have been together he has never done anything like this for me. Cooking dinner is one thing, but sending it to the office raises the game to a totally different playing field. He has really gone all out to show you how much he cares about you—and without sex being involved, I would like to add."

"I've never had anyone treat me this way before."

"Me neither. Pass the bread."

"I'm excited, but I'm also frightened."

"Frightened of what?" Sheila asked with a mouthful of bread.

"I'm afraid I might do something to run him away," I confessed.

"Tamara, just continue to be yourself. Because that is who Samuel loves. He simply loves you, not your body, your house, your car, your money, your achievements; just you. Love is a beautiful thing and it's not to be feared. So don't run from it or try to hide: just embrace it."

"You're right, as always. Thanks Sheila." I leaned over and gave her a hug. "Leon is very blessed to have you. You are a phenomenal woman and a great friend. I love you."

"I love you too. And I really love mister lunch-sending Samuel," Sheila chuckled.

We spent the remainder of the hour laughing, talking, eating and sharing like two school girls. Just as we were finishing up the snake appeared at the doorway. Clyde definitely was not going away quietly!

Chapter Thirty-Four
Sheila

"Good morning ladies!" Clyde beamed at us.

"Morning," I responded. Tamara didn't open her mouth to say a word.

"I'm sorry to interrupt. But Sheila, may I speak to you for a minute?"

I turned to Tamara and she looked as though she was about to attack Clyde at any moment.

"Please?" Clyde asked. "I promise not to keep you long." he said with his killer smile.

"Sure," I said. Tamara rolled her eyes and let out a loud sigh.

"Good to see you, Tamara," Clyde said as I walked out the door with him.

"You too, Reverend," Tamara said sarcastically with an emphasis on the word "Reverend".

As soon as we were a distance away Clyde started to plead his case.

"I can't believe you're treating me this way, Sheila."

"What way?"

"Ignoring me, that's what way! You act as though I meant nothing to you."

"Clyde, please."

"Is that it, Sheila? I meant nothing to you?" He was making me uncomfortable. I started fumbling with papers on my desk. "Sheila, look me in my eyes and tell me I meant nothing to you."

I looked up and stared Clyde directly into his eyes.

"We meant nothing to one another. It was just sex." I said to Clyde trying to discourage him.

"Maybe it was just sex for you," he said as he grabbed me.

"You had better let go of my arm!" I said jerking free of him. You listen to me and you listen well. You and I are no more."

"We'll just see about that." He reached over to kiss me, but I turned my head. He pecked me on the cheek and then walked away.

"I'll be talking to you soon, and by the way, please tell Leon I said 'Hello'." he said over his shoulder.

A cold chill ran through my body. *What in God's name have I gotten myself into?*

"So, what did that clown have to say?" Tamara asked. I was so shook up from Clyde's statement that I didn't hear her walk up behind me.

"That he still wants me, basically."

"You realize this is all about control for him. He can't stand the fact that you are ending things before he could get the chance to."

"How could I not see how crazy he was?"

"Believe me, he's not crazy. But he is possessive and with a man like that you never know what he will do. That's why I say you might want to reconsider telling Leon."

"Now you sound crazy!"

"Would you rather Clyde tell him?"

"You really think he would?"

"I wouldn't put anything past Clyde. I think that at this point he would threaten you with anything to keep you from leaving his clutches.

"This is turning into one big nightmare."

"Sometimes it doesn't pay to play. The cost is too high."

"Tamara, my life is unraveling right in front of me. I can't let Leon find out what I've done. He would never forgive me."

"Sheila, Leon loves you. He will be angry and hurt in the beginning, but I'm certain he will get over it. The two of you have too much history together to let it go because of one mistake."

"I'm scared, Tamara."

"As you should be. But I think now is the time for you to take your own advice and pray."

I hugged Tamara and didn't want to let go. Then I did exactly as she suggested.

Lord, I silently ask you to lead me through this time in my life and I promise to follow. Make things whole again. Build up what Satan is trying to tear down. Open Leon's heart to forgive me in Jesus' name. Amen.

Chapter Thirty-Five
Tamara

I left a little early so that I could have enough time to prepare dinner for Samuel. I was going to show him once and for all that I was very capable of cooking. He had been so good to me that I wanted to repay him for his thoughtfulness. It's not everyday that I get in the kitchen and rattle some pots for a man. But Samuel wasn't just any man, he was special. So the thought of coming home and sweating in a hot kitchen didn't bother me a bit. When I phoned him at work to tell him to be at my place around 6:00 tonight, he was curious as to why. But I just told him that I had a wonderful surprise in store for him. I hope I still knew how to turn the oven on!

Samuel arrived at 6:00 on the dot. I opened the door and greeted him with a kiss.

"Mmmm…it smells good in here. What have you been up to woman?"

"Just a little sumthin' sumthin' I threw together for my man. Right this way please." I said with sweetness. I led him into the dining room where I had set the table for two.

"So what do we have here?"

"Oven fried chicken, squash casserole and a fresh tossed salad."

"I'm impressed," Samuel commented.

"Have a seat."

"I don't believe this. Did you cook all of this yourself?" Samuel asked looking amazed.

"You better believe I did!" I sat down, picked up the platter of chicken and placed a piece onto his plate. I then, scooped up a heaping spoonful of the squash casserole and placed it next to his crispy chicken.

"Oh no! My rolls! They are still in the oven. Help yourself to the salad," I said as I jumped out of my seat and ran into the kitchen. I returned with a platter of hot buttered dinner rolls in hand. Fortunately, they were golden brown and not burned.

"Would you like a roll? They're hot right out the oven."

"Yes, I would love one." I placed a roll onto his plate.

"Dig in."

"If you don't mind I would like to say grace first," Samuel said.

"Of course."

"Father, we thank you for this food we are about to receive. May it nourish not only our body, but also our soul. Bless the hands that prepared it in Jesus' name. Amen."

"Amen."

Samuel took a bite of his chicken and closed his eyes as he chewed.

"Well, do you like it?"

"I love it! It tastes almost better than my mama's."

"Quit playing."

"I'm not. It's really good." He dug into the squash casserole. "Mmm."

"You need to stop."

"Woman, that's it! We're getting married. What are you doing tomorrow?"

"You are so silly!"

"Beauty, brains and can cook to boot. I've got to make you my wife!"

"So is that all it takes?"

"Well there is one other thing and this is extremely important. The answer will definitely determine if we are destined for each other."

"And what's that?" I leaned in closer with wide eyes to better hear the question.

"Do you snore?"

"Man, I thought you were being for real!"

"I am!" he said with a serious look on his face, but then burst out laughing.

We continued to laugh and talk until there was nothing left on our plates.

"Baby, that sure was good." He announced with a satisfied sigh.

"Better than the meal Sheila prepared?"

"The company was."

"Good answer. Let's sit out on the balcony for awhile."

"After you." Samuel bowed and extended his arm for me to lead the way. The sun was just starting to set, creating a serene atmosphere for the moon to appear.

"What a beautiful view," I said to Samuel.

"Yes it is," Samuel commented. But as I turned to face him he wasn't looking at the view. He was looking directly at me.

"I was referring to the sunset."

"But I'm referring to you. You're beautiful Tamara and everyday I find a new reason to love you even more." With that he kissed me.

"You're shivering. Are you cold?" Samuel asked.

"No, you just overwhelm me. I've never met anyone like you. I never knew anyone like you existed, and now that I have found you I never want us to end."

"Sweetheart you don't have to worry about that, because I don't ever plan on going anywhere. You are the only woman for me."

"How can you be so sure?"

"Because from the moment I laid eyes on you my heart has never beaten the same."

"You are so full of it," I teased him.

"I'm full of love for you baby. I would run through hell with gasoline drawers on for you."

"Boy, you are crazy."

"I'm not crazy, I'm just a man in love." He kissed me again.

All of this kissing was driving me crazy! He had better stop while he was ahead, because I was at the point of ripping his clothes off right there on the balcony for the world to see. The tension between us was building and he pulled away.

"So when do I get to meet your parents?" he asked suddenly.

"My parents?"

"Yeah, I've always heard if you want to really know the person you are going to marry you need to meet their parents."

"Marry? You really want to marry me?" *He actually wants me to be his wife.*

"Woman, have you been listening to anything I've said to

you? I love you. Don't you love me too?"

"Yes, you know I do."

"How am I supposed to know? You don't tell me very often, only when I say it to you. Do you have a problem with saying it?"

"Of course not."

"Then tell me you love me."

"I love you. There I said it. Happy now?"

"Not particularly."

"Good grief I said it. What more do you want?"

"For you to mean it. But I guess you don't feel it as I do."

"I do mean it Samuel. I merely don't express it the same way as you do. But that doesn't imply that I don't love you just as much or more. I've told you all of this is new to me. I'm not accustomed to cooking dinner for a man and taking the time to get to know him, or him being genuinely interested in getting to know me. I've never spent this much time talking and expressing with a man other than in a horizontal position. So please be patient with me."

"I apologize baby. I was looking at things from my own selfish standpoint. I wasn't considering the fact that you cooking that fabulous dinner was an expression of your love for me. Do you forgive me?"

"How can I not?"

"That's my girl," Samuel said holding me tight.

"Samuel."

"Yes dear."

"I love you."

"I love you too." He held me tighter. I was in heaven and could have stayed in Samuel's arms all night. I never knew love could feel so good.

"Tamara, I've wanted to ask you something and I can't let this opportunity pass me by any longer," Samuel whispered in my ear.

Oh goodness, I've heard that before. Now I've already gone to church with him, so what else could it be…a tent revival? Samuel got down on one knee.

"I know we haven't known each other long, but I know in

my heart that you are the woman for me. Forget meeting your parents. Would you please make me the happiest man in the world and agree to be my wife?"

"Samuel, I don't know what to say?"

"Say 'Yes'!"

"But are you sure?"

"Woman, I've never been more sure about anything in my life. Will you be my wife?"

"Yes! Yes, I'll be your wife!" I was crying so much that I couldn't see the ring that Samuel was showing me. He grabbed my hand and slipped it on my finger. That's when I totally lost it and started sobbing uncontrollably. *How I wish my mom was here to experience the happiest day of my life with me.*

Chapter Thirty-Six
Sheila

When I arrived home from work, Leon had gotten home early and was playing Play Station 3 with Kelsey. I loved it when they were having their father and son time, so I didn't bother them. I went straight to the kitchen to prepare dinner along with my sidekick. Indya loved helping me with dinner, or any chores for that matter. Seeing that I had experienced such a stressful day, I decided on something quick and simple. While I browned the ground beef for the spaghetti, Indya stirred the sauce. We were a great team, but Indya was a chatterbox and today I was deep in thought. I had managed to dodge Clyde on his two attempts to engage me in conversation. But I was concerned about what he would try next to gain my attention.

"Mommy, when I grow up I want to be a mommy just like you," Indya said bringing me out of my daze.

"But what about college?" I asked her.

"They teach you how to be a mommy in college?" Her innocence was so sweet. *If only she could stay that way forever.*

"No, honey. Being a mommy comes natural. You go to college to learn other things such as how to be a doctor or lawyer."

"But I don't want to be one of those. I want to be like you."

Be like me and what exactly is that? An adulteress, a phony and a hypocrite, because that's what I felt like today. This secret is beginning to really mess with me. Maybe telling Leon is the best thing to do.

The spaghetti was done and I was taking the garlic bread out of the oven. My little helper had tossed together a salad, which consisted of her opening up a bag of greens and placing them in a bowl. She was so proud of herself.

"Go tell your father and Kelsey that dinner is ready, baby."

"Okay, Mommy. Daddy, Kelsey dinner's ready!" Indya screamed from the kitchen. I could have done that.

The two guys came running into the kitchen and took their seats at the table.

"Smells like mommy's famous spaghetti. All right!" Kelsey proclaimed.

"I stirred the sauce," Indya proudly announced.

"Then I bet its extra good," Leon answered her.

She was grinning from ear to ear. She was definitely a daddy's girl. We ate our dinner while Leon and Kelsey bragged on who beat who in which game. They both were really competitive and hated to lose. It was an event just to hear them relive the games all over again.

"Would anyone like some more spaghetti or salad?" I asked.

"Not me. I'm stuffed," Leon said patting his stomach.

"Me, too," Kelsey said mimicking his father by rubbing his stomach.

"I'm still eating Mommy," Indya said with a mouthful of food.

"Then I'll start on the dishes," I said getting up from the table.

"No, honey. Kelsey and I will do the dishes tonight."

"What?" I asked with surprise.

"What!" Kelsey exclaimed.

"You've done enough for today. Kelsey and I can at least do the dishes," Leon said staring at Kelsey, daring him to say anything to object.

"Aww man," Kelsey sucked his teeth. Leon reached over and thumped him on the ear. "Ouch!"

"Why don't you go upstairs and take yourself a bubble bath? I'll be up later to wash your back." he winked at me.

I know what Mr. Leon was up to. He wanted some grown folks time. I can't blame the man for that. I was just happy to get out of doing the dishes and grateful for a diversion from what was really on my mind.

I settled into the tub listening to the laughter coming from downstairs and smiling to myself. *God has blessed me with a wonderful family. I'll be darn if I was going to allow the devil to come in between what the Lord had so graciously given to me.* I had closed my eyes and was just starting to relax when I heard Leon walk in.

"Can I wash something for you, my dear?"

"No, but you can join me."

"I don't mind if I do," Leon said stripping off his clothes.

He crawled into our Jacuzzi tub and I rested my head on his chest.

"It's been a long time since we have bathed together. Why did we stop?" I asked Leon.

"I don't know. I guess we just stopped making the time. But all that is about to change." He wrapped his arms around me.

"Do you promise to love me forever?"

"For an eternity," Leon answered me kissing the back of my head.

The phone rang disturbing our vibe.

"Don't you move. I'll get it," I said grabbing a towel. I answered the phone after about five rings. Whoever it was they were persistent.

"Hello."

"Good evening, Sheila." My heart was racing. I couldn't believe he had the audacity to call my house.

"How dare you call my home. You have gone too far Clyde," I whispered into the phone.

"Well, you wouldn't talk to me today when I tried to. So, what do you expect me to do?"

"Clyde, please don't do this. I'll do anything," I pleaded.

"Then meet with me."

"Fine."

"I'll see you tomorrow my darling."

I hung up the phone and just stood in place unable to move.

"Honey, the water's getting cold," Leon hollered from the bathroom.

I had two options, tell my husband my ex-lover had just called me or go along with the masquerade and meet Clyde on tomorrow. I walked back into the bathroom trembling with fear.

"Who was that on the phone?" Leon asked.

"Oh, it was a wrong number," I lied.

Chapter Thirty-Seven
Tamara

In spite of getting very little sleep, I was full of energy the next morning at work. I was so excited I couldn't even speak when I approached Sheila, so I just shoved my hand in her face. The look on her face was priceless. Her mouth flew open.

"No, you didn't!" Sheila screamed.

All I could do was nod my head.

"You're getting married?" She screeched.

Sheila sprung from behind her desk and hugged me until I couldn't breathe.

"I am so happy for you. So, how did he propose? I want to know everything and you better not leave out one detail!"

I told Sheila about the entire evening starting with my tasty dinner until the moment Samuel asked me to be his wife. We talked like two schoolgirls forgetting the fact that we were at work. In just a few short hours we had the whole wedding planned out.

"You know I can't do any of this without you," I said to Sheila.

"Of course you can! But you won't, because I wouldn't let you. I've been waiting on this day since the day after I met you!"

"You are such a mama!"

"Speaking of mamas, are you going to call and tell yours about your big news?"

"And what am I supposed to say? 'Hi Mom. I know it's been a few years but I just wanted you to know that I'm getting married'."

"That would be a start."

"Sheila, please. I don't think so."

"But you can't just not invite her. She's your mother."

"I know who she is. I don't need you to tell me that." I said with more of an edge in my voice than I intended.

"Tamara, I realize that you and your mother's relationship is a very touchy subject and I didn't mean you any harm. I only wanted your very special day to include her. It's not every day that a girl gets married." I rolled my eyes as Sheila spoke. "Okay, I'm not going to preach to you. But just think about it before you discard the notion."

"I hear what you're saying and I appreciate it. It's just been so long…"

"Too long. Let the past stay in the past and call her."

"You want to know something weird? One side of me wants to call her but the other side is still hurt. My first thought after Samuel proposed was how I wished my mom was here. I was crying because I was ecstatic that Samuel found me worthy enough to be his wife, but I also was crying out for her. Truth be told, I really do miss my mom. I've just learned to do without her in my life. But its times like this that I wish she was near."

"Then call her. Today is a good day."

"I don't know. She may not want to hear from me."

"Girl, please! That's your mother, the woman who birthed you into the world. Take it from me she will be delighted to hear from you. She has probably been praying for that day to come."

"Then why hasn't she called me?"

"Don't go there, Tamara. We could ask why all day long and still not come up with an answer. Why doesn't matter? It's when. When are you going to drop your pride and pick up the phone?"

I didn't have the answer to Sheila's question. But I did know that I had better come up with an answer quick because time was running out. I had a wedding to plan and how would it look if the mother of the bride weren't on the guest list.

Chapter Thirty-Eight
Sheila

After I had finished talking with Tamara about her wedding plans, I turned my attention back to work. I checked my voice mails and missed a call from Clyde, which in my opinion wasn't a bad thing. He wanted to know when I was going to be free today. I dreaded calling him back, but I had agreed to meet with him. I'll call him later.

"What are you doing for lunch today?" Tamara asked me from the doorway of her office.

"I'm meeting my hubby at The Willow."

"How nice."

"What about you?"

"I was going to run over to the mall and check out some wedding gowns."

"Oh, fun. I want to go. You can't wait on me?"

"That was my reason for asking you about your lunch plans today. But not to worry, this will not be my only trip to the mall. I have a feeling that by the time all of this is over we both will be sick of the place!"

"You're right! But try not to have too much fun without me this time." Just as I got the words out Leon was stepping off the elevator. It should be a sin what the man does to a pair of Levi jeans.

"Honey! I thought we were going to meet each other at the restaurant?"

"I know, but I decided that I'll make things easier for you by picking you up," Leon said kissing me on the cheek.

"Hi, Leon!" Tamara said.

"Hey, Tamara. I didn't see you standing over there, girl. How's Samuel?" He walked over to Tamara and gave her a hug.

"He's wonderful," Tamara answered as she waved her hand in Leon's face.

"What is this? You mean to tell me that Tamara done gone and got herself a husband." Leon grabbed Tamara's hand to get a closer look. "Congratulations."

Leon and I left for lunch. The Willow was on the other

side of town and I was glad Leon had come to pick me up. It took us nearly thirty minutes to get there.

"I can't believe Tamara is getting married. I didn't know her and Samuel was that serious."

"I saw it coming. I couldn't be happier for her," I said.

"How long have they been dating?"

"Not even two months."

"Get out of here!"

"You sound as though love can't happen in that length of time. You know I had your heart after our first date."

"You mean I had you! You were putty in my hands."

"Man, you were a permanent fixture in my living room. I was about to tell you to chip in on the rent, but you proposed instead."

"Do you ever regret it?" Leon asked turning serious.

"Not a day."

"Together forever," Leon said placing his hand on mine.

At that moment I decided I wouldn't be meeting with Clyde and jeopardizing the love I shared with my husband. Tamara was right. We had history and a lifetime of memories ahead. Our relationship wasn't perfect, but one thing was for sure, we loved one another and that was worth hanging on to.

The rest of the workday was pretty uneventful except for Tamara's excessive chatter about the wedding dresses she had tried on at the mall. Listening to her brought back memories of my own wedding day. It was one of the happiest days of my life. I was determined to get back to that day.

When I arrived home, Leon's car was unexpectedly in the garage and the children were nowhere in sight. He must have gotten off early to surprise me with something special. I couldn't wait to find out. I found him in our bedroom standing in front of the bay window.

"Hi, sweetie. What brings you home so early?" I asked him smiling.

He turned around and the look on Leon's face let me know something was wrong—terribly wrong.

"Did you sleep with Reverend Peterson?"

I was caught off guard and couldn't speak.

"Don't try to lie. He told me everything, every sickening detail!"

"Leon, I can explain."

"Explain? How do you explain it?" he yelled.

"I didn't mean for it to happen."

"What the hell do you mean you didn't mean for it to happen?" Leon was standing in my face. "You had a choice!"

"It was a mistake."

"No, we are a mistake! All of these years I loved you with all of my heart. I would do anything for you and this is the thanks I get. You cheat on me! You lay with another man! The preacher of our church—of all people! Sheila, I never thought you would betray me like this." Leon started to cry. I had never seen him shed a tear before, not even at funerals.

"I'm sorry! I never meant to hurt you." I reached for him.

"Don't touch me!"

"Leon, I love you."

"You love me? You have no idea what love is. Because if you did you would have never been with another man. You're nothing but a selfish, spoiled brat who thinks the world and everybody in it is supposed to revolve around you. Well, I'm tired Sheila and I'm done. As far as I am concerned, I hope you and 'the Reverend' are very happy together. I'm outta here."

"Leon, please don't go!" But he only pushed past me.

"I'll be by this weekend for my things!"

"But where are you going?"

"Anywhere, but here."

"Leon, let's please talk about this," I ran behind him tugging at his shirt.

"It's too late to talk. I can't even stand the sight of you right now."

"But you can't really want to throw everything away, all that we have, the years we shared."

Leon stopped in mid stride and turned to face me. "I didn't, you did. The moment you stepped outside of our marriage. Throughout all of our problems I never turned to another woman. I never wanted anyone, but you. You were my life, Sheila. You and the kids are all I have." Now we both were crying. I wanted to hug him, but he wouldn't let me get close enough to him.

"How could you do this us?" He walked out of the house.

"Leon, don't leave me!" I screamed after him, but it was too late. He was gone. My worst nightmare had come to life.

Chapter Thirty-Nine
Tamara

My mind was swimming with visions of all the gowns I had seen earlier today. There were so many to choose from. I tried on one with spaghetti straps and a fishtail skirt; a satin A-line with a beaded overlay that was as smooth as butter against my skin; a striking v-neck halter; and yet another one with a mermaid cut that hugged my body and flared out below the knee. They all were so beautiful and just when I thought I had found "the one" the sales lady would bring out another that was as beautiful. I definitely needed Sheila's help to select the right dress just for me. What fun we were going to have. Special days like these were designed for special friends…and mothers.

Ever since Samuel had proposed, I thought of my mother and how I wanted her to share in this celebration with me. I had graduated from college without her, bought my first home without her, landed my job and got promoted without her. I couldn't get married and start a family without her being there too. Life wouldn't be complete.

My cell phone rang.

"Hello."

"Tamara, Leon's gone! He's left me!" Sheila screamed into the phone.

"What are you talking about? Slow down."

"He found out and he's left me. He's gone, Tamara. What am I going to do?" Sheila was hysterical and I had to get her to calm down.

"Sheila, tell me what happened. But first I need for you to get control of yourself."

"Clyde told Leon everything. He asked me to meet with him today. When I didn't he called our home and told Leon about the affair," Sheila sobbed.

I could barely understand what she was saying because she was talking and crying at the same time. But I did hear that Clyde was behind it all. "Where are you, Sheila?"

"I'm at home."

"I am on my way," I told her. I was so mad that I was

seeing red. *I personally want to whip Clyde's behind. He needs to be stopped.* I made a quick call to Samuel and explained to him what was going on. He said he would drive over to Leon's shop to see if he was there. I hoped that he was and that Samuel would be able to talk some sense into him or at least knock him unconscious and drag him back home. Whatever worked. Desperate times calls for desperate measures.

I arrived to Sheila's house in record time. The front door was wide open. I ran inside. She was pacing the floor and looking like a crazy woman. She had cried until her mascara had run down her face, her blouse was hanging out in the front, she had one shoe on and the other off. She was really out of it. She didn't even hear me come in.

"Sheila." I called to her. She stopped in her tracks and hobbled over to me. I met her with a hug. I didn't know whether to be mad with her or pity her, so I just held her and let her cry in my embrace. "I'm so sorry. But we are going to get through this, everything is going to be fine," I tried to console her.

"No, it's not. He will never forgive me and I can't blame him. I don't know what I was thinking. I ruined everything. Nothing will ever be the same."

"No, it won't be the same, but that doesn't mean it can't be better. You made a mistake, sweetie. People were hurt, but no one died. So you still have a second chance to turn things around."

"But you didn't see the look on his face or hear the way he talked to me. I'm afraid I have hurt him beyond repair. My life is over."

"Your life has been altered, but it's far from over. Where are the kids?"

"The kids? I don't know. They weren't here when I got home."

"You mean they're still at school?"

"No, Leon probably took them to Mama Lyn's so they wouldn't bear witness to when he confronted me. Mama Lyn! How am I going to tell her Leon left me because I slept with another man? Not just a man, but Reverend Peterson! What am I going to tell the children?"

"You're not going to tell them anything just yet. I have

Samuel looking for Leon as we speak. Any ideas on where he might be?"

"The shop would be my first guess."

"He was already on his way heading in that direction. Let me try him on his cell and see if he had any luck." I flipped out my cell phone and dialed Samuel's number. I stepped to the kitchen to talk with him.

"Samuel just left the shop and Leon wasn't there. He said the guys told him he received a phone call earlier and left soon after. They haven't heard or seen him since. Any other ideas?" I asked Sheila.

"You don't think he would go directly to see Clyde, do you?"

"Lord, I hope not. Does he know where Clyde lives?"

"He doesn't have to, because he knows where the church is. Tonight prayer service is being held and Clyde will be there."

"We better get to church. I'll call Samuel and have him to meet us there."

How ironic that we were all heading to church on the night of prayer service when prayer was most definitely in order, maybe this is God's way of getting everyone together in His house at the same time, on one accord. I hope we aren't too late. I know one thing: after we get out of this mess, I am going to sit Samuel down and tell him everything there is to know about me.

I would never admit it to Sheila, but I was more afraid than she was! I was scared of what Leon would do to Clyde if he saw him face to face. Because one thing you don't do is mess with another man's woman. Clyde doesn't know it, but he was close to having a near-death experience.

Just as I dreaded, Leon's truck was parked in front of the church when we pulled in, along with Clyde's Lexus and a few of the parishioners' cars. Sheila was already crying. I wanted to cry too, because Samuel hadn't made it yet. Now what were we going to do?

I thought I had seen some movement in Leon's truck. He hadn't gone inside. It was up to me to talk to him. Sheila sure wasn't in any condition to do so. Just as I opened my door to walk over to Leon, Samuel drove up.

Thank goodness! Making eye contact with Samuel I pointed

in the direction of Leon's truck and he parked right next to him. He got out of his car and slid into the passenger side of Leon's vehicle. I quickly walked back to my car before my legs collapsed from up under me. Sheila was still crying. We sat in silence and waited. My thoughts drifted back to the stunning wedding gowns that I had tried on earlier. To think it had started out as a good day. My cell phone rang, startling me.

"Hello."

"Hey, baby. You girls can go. Leon and I are going to drive around and talk."

"Is he okay?"

"He just needs to let off some steam. I'll call you later."

"I love you."

"Back at cha." We hung up.

"Samuel's here and has everything under control. He told me we could go home."

"Then that's just what I want to do. I want to go home and wait for my husband."

"What about the kids?"

"Oh yeah, the kids. You feel like taking a ride to Mama Lyn's?"

"Are you going to tell her?"

"I'm not going to have to. She is going to read it all over my face."

"Are you nervous?"

"Petrified."

"What do you think she is going to do?"

"Rebuke this lust demon right out of me."

"Well if that's the case, you should have gone to her in the first place."

Chapter Forty
Samuel

I had only met Leon a few weeks ago, therefore, I didn't know him very well, but I think after this encounter we were going to be a little closer. Though it would require us getting personal and honest with one another. Self-disclosure is something very foreign for men, because as boys we were raised to hide our true feelings and were told "Real men don't cry". But thank God my father was the opposite and didn't buy into that. He encouraged me to express myself and to always be in tune with my emotions. Afterall, even Jesus wept.

"I'm here to help in any way I can, Leon," I said to him as I slid into the passenger side of his truck.

"The only person who can help me is the man inside that church," Leon said angrily.

"I know you probably don't want to hear this, but there is another man."

"Who? You?" Leon looked at me for an answer.

"No, Jesus."

"Man, don't come at me with that religious mumbo jumbo. Haven't you heard it was a preacher who slept with my wife?"

"A preacher did, not Jesus. Don't let one depraved man keep you from getting what you need. Jesus is the only one who knows just how badly you're hurting and He's the only one who can heal you. Don't turn your back on Him."

"I'm too angry. I'm angry with Reverend Peterson for deceiving me. I'm angry at my wife…" Leon lowered his head. "How could she…?" He broke down and begun to cry.

"Let's take a ride. I'll drive." I called Tamara to tell her to take Sheila home, because Leon and I had some male bonding to do. We then switched sides so that I was the one driving his truck. He was in no state to be behind the wheel.

"Leon, it's going to be alright. I know it doesn't seem like it now, but things will get better. Your marriage can survive this. You just have to take it one day at a time."

"I don't know if I want it to. How can I trust her again? How can I ever…?" He broke down again.

Leon was so hurt and angry at this point he wasn't hearing anything I had to say. I decided to pull over into an empty lot. Before we went any further I needed to pray.

"Leon, I want to pray for you," I said to him and switched the ignition off.

"Whatever man."

"Oh Heavenly Father, if ever a time we need you, we need you now. Lord, hearts have been broken, feelings hurt, promises shattered and faith challenged. But where love and trust have diminished I ask that you restore it right now. Father, I ask that you take away anything or anyone that could pose a threat to Leon and Sheila's marriage. Sever all unholy ties. Establish a bond of unity between them that can not be destroyed. Give them the strength to hold on to all that is good in their marriage. Teach them to seek your face in the midst of trouble. May they be like-minded in making decisions concerning their family and their future. Shield them with your hedge of protection and may they be a source of happiness for each other for years to come. Bless their union Father and I count it done in Jesus name. Amen."

By the time I finished with my prayer Leon was really letting it out, so I reached over and gave him a brotherly hug. He embraced me back tightly as though his life depended on it. I knew exactly how he felt, because not too long ago I had been there. Only love can make the biggest man cry like a baby. You feel helpless and there's no escaping the pain.

"It's going to be alright, my brother," I said, trying to console him.

"I hope you don't think I'm some kind of wimp, man," Leon said wiping his eyes.

"Brother please! Don't ever think that. Feel free to let your emotions flow. Your world has been turned upside down. You feel betrayed. If you weren't affected by all of this, then I would have reason to worry. Your tears aren't wimpy, they're natural."

"I just don't know what to do or which way to turn," Leon said shaking his head. "I'm lost. Where do I go from here?

"Back home to your wife to patch things up."

"I can't see doing that, man. I don't think I can stand looking at Sheila. Do you know how badly I wanted to wring her neck today? I had to get out of the house, because I kept envisioning my hands around her skinny little neck. Damn, I'm mad!" He hit the dashboard of the truck.

"One thing my father used to always say and I have found to be true is that time brings about a change. All of the emotions that you are feeling today will disappear. There will be a day when you'll be able to look at Sheila with love in your eyes again."

"I don't know about all that, man."

"You can get beyond this point. But first you have to find it in your heart to forgive Sheila."

"I want to, but I don't know if I can."

"Just think about the day you married her. Remember how she made you feel and how in love you were? Get back to that day." Leon didn't respond, but just nodded his head. "Aside from this one incident Sheila has been a good wife. I'm not trivializing what she's done. I understand the disappointment that goes along with her indiscretion. But know that out of our humanness we sometimes make mistakes and bad judgements. But that doesn't negate the good that's in us. Think of the years, the memories and the dreams you two have together. Your marriage deserves a second chance."

"But how?"

"With the help of the Lord. He can fix it just like new. He did it for me and He will do it for you."

"You've been married before?"

"No, but I came close. My fiancé ran out on me and to this day I don't know whether she is alive or dead. I never thought I would find love again. It really tore me up inside. I closed my heart to the possibility of ever loving again. But then I met Tamara and it honestly was love at first sight. I thank God for her each day. Because before her I wasn't living, but merely existing. Now life is better than ever all because I opened my heart to love and it can happen for you too. All you have to do is open your heart, brother."

"I do love Sheila with all my heart. I've never loved any woman like I've loved her and I truly couldn't imagine my life without her. I know I'm a better man because of her. But I never thought she would hurt me so badly."

"I understand your pain, but I'm here to tell you that this

too shall pass. The best way to get through it is to talk it through. So what you need to do is go home and talk to your wife." Leon didn't answer. He continued to gaze ahead. "Do you think you are you ready for that?"

He let out a sign. "I think I'm gonna need a beer first."

"A drink is the last thing that you need. Get drunk on the word. That way you don't have to worry about waking up the next day with a hangover."

"You know, you're alright, Samuel," Leon said smiling slightly. "I guess I could try, but I'm not promising you anything. But I do need some answers right now."

"Well that's all I ask is that you try, that's all anyone can ask." I switched the truck's ignition back on and began to drive off.

"Hey man, thanks. I appreciate everything you've said and done. It really means a lot and I will never forget it."

"No need to thank me. I was only doing what brothers do for one another."

"Well I appreciate you…brother."

"Anytime." *Thank you Lord.*

"Oh yeah, I hear congratulations are in order for you. I wish you and Tamara the very best. She's found herself a good one."

"Thank you. We both are blessed to have each other, just as you and Sheila are a blessed union as well. What God has put together let no man—not even a preacher man—separate! The devil knows you are blessed and he is trying his best to destroy what you have. Don't let him. Your family is too important."

"Well he has messed with the wrong man now. I love a good fight."

"That's what I'm talkin' bout."

Chapter Forty-One
Sheila

The thought of facing Mama Lyn was agonizing, but I had made my bed, now I had to lie in it. When we arrived at Mama Lyn's, it was as though she was waiting for us. We found her sitting at the kitchen table.

"Hey Mama Lyn," I greeted her with a kiss on the cheek.

"Hi babies. Y'all have a seat."

"Where are the kids?" I asked nervously.

"The kids are fine," Mama Lyn answered.

"I hope they haven't been any trouble," I said stalling.

"You know those children haven't been no trouble. Now, you stop beating around the bush and tell me what in the world is going on." There were a few minutes of silence while I searched for just the right words to say.

"Leon's left me," I finally spoke.

"And why would he do something like that?" Mama Lyn asked. She was not going to make this easy.

"He found out that…" I couldn't bring myself to tell her. I couldn't say it.

"You might as well just spit it out," Mama Lyn ordered.

"He found out that I had slept with Clyde, Reverend Peterson," I blurted out.

"Lord Jesus!" Mama Lyn just shook her head. She stood up from the chair she was sitting in and I feared she was going to strike me. But she walked past me. She was shaking her head and rubbing her hands together. Her quietness was making me uncomfortable, so I went to try to explain.

"I didn't mean for it to happen."

"So, why did it? At some point before you crossed the line you had a choice. You had a choice to walk away, but you chose not to. In a split second you made a decision that jeopardized your marriage and your family. Well, was it worth it?"

"No ma'am." I answered tearfully.

"It's too late for tears, Sheila. I'm not buying it."

"But I'm sorry," I sobbed.

"Everyone always is, after the fact! But after the fact is too late! I don't care what kind of problem you and Leon were having or what you thought he was doing, he didn't deserve this. What were you thinking chile?"

"I wasn't thinking Mama Lyn. I was just lonely and Clyde was there."

"Then you should have gone to your husband and not Reverend Peterson! Reverend Peterson! Lord have mercy!" Mama Lyn said still shaking her head at me in disgust. "Why didn't you come to me?" she asked as she sat back down in her seat next to me.

"I was so ashamed."

"Ashamed of what?"

"I was ashamed of the martial problems me and Leon was having. I thought after of all the years we've been married we should have been past all the petty stuff. I was ashamed that I let our problems cloud my judgment and it caused me to have feelings for another man. How could I tell you that, let alone God?"

"Don't you ever think you can't come talk to me about anything. I am a woman and can understand what you've going through. Remember I've been married, too, God rest your grandfather's soul and I know how hard it is and the work it takes to keep a marriage together. But I also know you don't have to do it alone. We have Jesus as our intercessor. We can go to him when we think we can't go to anyone else. And sweetie what you have been feeling is nothing new up under the sun. The devil wants you to think that you are all alone and there is no hope. That's how he creeps in. He used you just like he used Eve in the garden to take down the head. That's who he's really after. Leon. As a matter of fact, call Leon and tell him Mama Lyn wants to talk with him right now."

I didn't think Leon wanted to hear from me, so I asked Tamara to call Samuel and have him to relay Mama Lyn's message. The truth was I was embarrassed, and I wasn't ready to face him and see the pain in his eyes—the pain I caused. As we sat and waited for Leon to arrive, Tamara told Mama Lyn that she had just recently become engaged. She told Mama Lyn all about Samuel, and I could tell how Mama Lyn was smiling and nodding her head that she was impressed and happy for Tamara.

In just a short time Leon and Samuel walked into the kitchen. I was so nervous my palms were clammy. Leon went directly to Mama Lyn and fell into her arms. He buried his head into her chest and cried. Samuel walked towards Tamara and they held each other. I lowered my head because I was left alone. But I felt Mama Lyn pull me up and I was standing in front of Leon. She pushed us together and I was surprised to feel Leon embrace me. We held each other and cried. I whispered to him how sorry I was. Mama Lyn instructed Samuel and Tamara to surround us and hold hands. Then Mama Lyn began to talk to the Father.

"Lord, we come before you standing in the gap for Leon and Sheila. We ask you for your forgiveness Father. We also ask that you enable Leon and Sheila to forgive themselves, as well as, each other and to not look back after this day. Create in them a clean heart and renew in them a right spirit. Where pain and anger reside, replace it with love. Heal their wounds, Father and bind any thoughts of divorce and separation. What the devil meant for bad make the ordeal be for their good and draw them closer together as never before. I speak blessings upon this marriage. Bless them as they seek your will in the mighty name of Jesus. Amen. Thank you, Jesus."

I couldn't see Mama Lyn, but I could still hear her calling on the name of the Lord. She then started to pray for Samuel and Tamara. Samuel became so full of joy that he started speaking in tongues. Tamara too was crying uncontrollably and thanking Jesus. Leon and I continued to hold on to each other. The spirit of the Lord had definitely come down and I knew none of our lives would ever be the same.

Chapter Forty-Two
Tamara

"That was really intense. I have never experienced anything like that before." I said to Samuel.

"Are you okay?"

"I'm better than okay. That was truly remarkable."

"Yeah, I think we all needed that."

"Do you think they are going to be alright?"

"They are going to be better than alright. They have the Lord on their side and Mama Lyn! And she ain't nothin' to play with."

The rest of the ride was quiet. After our praise session at Mama Lyn's, Leon and Sheila drove home together and I was driving Samuel back to the church to get his car. I was contemplating when would be a good time to tell Samuel about the skeletons in my closet.

"Honey, I think we need to have a talk."

"Is something wrong?"

"There are some things you don't know about me and I want to come clean."

"Some things like what? You haven't slept with the preacher too, have you?"

"Samuel, no! It has to do with my childhood and my parents." He gave me a quizzical look.

We had made it back to the church. Clyde's car was still there, so was another car, probably that of his latest victim.

"Why don't you follow me to Sugar and Spice? I could use a big chunk of chocolate cake."

"I'm right behind you."

It took us about fifteen minutes to drive to the café and no time to be seated. I ordered their three layered chocolate cake with a decaf coffee. Samuel ordered his usual French Vanilla latte and waited patiently for me to begin. I began with Uncle Ray.

During the course of telling my story, I had to pause several times because I was getting choked up. It was amazing to me that after so many years it still made me emotional. I guess that was a sign that I wasn't healed. Samuel was very supportive and held my hand the whole time. I explained to him how I felt that my father had abandoned us for another woman and I hadn't spoken to him in almost twenty years. I told him how my mother wasn't there for me when I confided in her regarding my uncle and how after high school I cut ties with her. It was as though Samuel was my priest and I was in a confessional. It was freeing to finally have everything out in the open.

After I finished, he squeezed my hand and told me that he loved me. I couldn't believe it. After hearing all of my awful secrets, he still loved me. All of this time I had been afraid of losing him if he found out about the real me.

"I'm sorry, baby. But I'm happy that you trusted me enough to tell me. I know it wasn't easy."

"No, it wasn't. But it feels better getting it out."

"So when are you going to call your parents and invite them to our wedding?"

"I'm not sure when, but I have been thinking about it."

"Well, stop thinking about it and just do it. Enough time has been wasted and tomorrow is not promised to us. Call them because I can't wait to meet my new in-laws." Samuel said with a smile.

"And I can't wait until they meet you."

"Well, I say let's get the ball rolling. Why don't we get out of here so that you can call them tonight?"

"Tonight? No, I can't. I don't think my senses could take much more tonight."

"If not now, when?"

"Soon, I promise."

Later on that night as I lay in my bed, I couldn't stop thinking about my parents. I had missed them over the years, especially my mom. Sleep was nowhere near for me, so I sat up in bed.

I thought about my life and how the harder I tried to run from the ghosts of my past the attempt became more tortuous. The heaviness of carrying around all of this unforgiveness in my heart was becoming too much to bear. I needed peace. It was time to let go of the bitterness I felt towards my uncle, the abandonment issues I had with my father and the resentment I was holding against my mother. I was ready to be set free.

I kneeled down on the side of my bed. I didn't have any idea as to what to say, so I spoke to God from my heart.

"Heavenly Father, I know it's been along time since you've heard from me and for that I'm deeply sorry. My heart's been so full of anger and I allowed it to blind me to your love. But I'm ready to come back home. Please forgive me for the sins I committed against you. Purge me from all unrighteousness and deliver my soul from this pit of despair. Release me from my past and the power it holds over me. Heal me from my brokenness. Lord, make me whole. I want to be set free to be the person you created me to be. Fill me with your Holy Spirit. I put all my trust in you. No more will I look to other people to complete me. My strength comes from you. You are my God, my refuge and my fortress. I surrender all in the precious name of your son, Jesus. Amen."

I stayed on my knees long after my prayer and cried for the last time over the things I had lost. After today my life was going to be different. Better. I was starting over with the Lord on my side.

I opened the drawer of the nightstand that was beside my bed and took out my address book. I found my mom's number.

My Aunt Josephine, my mom's youngest sister, always kept me abreast of how things were going in my mom's life. I knew that she still lived in South Carolina and had never remarried.

I picked up the phone and dialed her number. She answered on the third ring. Even though time and distance had separated us I still recognized her voice. It was music to my ears.

"Hello," She answered in a hazy voice.

"Hello," I whispered.

"Tamara? Is that you?"

Praise God. She recognized me, too. Life Lesson #3...Obedience is better than sacrifice.

Chapter Forty-Three
Sheila

On the way home I repeatedly told Leon how sorry I was and asked him to please forgive me. He simply nodded his head and patted me on the leg. Though he tried very hard not to show it I could still see the hurt in his eyes. I pray that I hadn't damaged him too badly.

After we arrived home, I instructed the children to take their baths and get ready for bed. I needed to prepare for work tomorrow myself. Leon was fortunate to have the day off. I suppose he would need the time to himself. No thanks to Clyde. I wasn't sure how I was going to handle him, whether it's best to ignore him or confront him. Maybe I'll make a call to the church and give them an earful regarding their beloved preacher. But that would only add fuel to the already lit fire and it isn't fair to put Leon through anymore heartache and hardship. Maybe we should have just let Leon have his way with Clyde. He would have at least felt better. I know Leon; his anger has to come out—one way or another. He was being too calm.

Leon stayed downstairs in the den as I went to check on the kids. They hadn't noticed anything different about the day, which was good. I wasn't ready to answer any questions or explain to them that I had messed up. After getting them settled, I started to run my own bath water. Considering I had just experienced the day from hell it was going to be nice to soak in the tub. *I still can't get over Clyde. He really had me fooled, but Tamara seen right through him from day one. He used me and my vulnerability for his own pleasure. I wondered how many women did he prey on and catch off guard? I wonder if his wife was aware of his deceitful behavior? He probably has her fooled, too.*

I finished my bath, dried off, and slipped into a nightie. Leon was still downstairs. I thought maybe he had fallen asleep, so I pulled the covers back down on the bed and went downstairs to locate him. I found him at the kitchen table sitting in the dark.

"Honey, are you coming to bed?"

"Not yet."

"Can I get you anything?" I said walking toward him.

"No, Sheila. I really want to be alone right now. Just go to bed without me."

I fought the urge to tell him once again how I was sorry. I didn't want to leave him, but at the same time I didn't want to aggravate him either. So I kissed him on the forehead and went back upstairs. Only time and prayer were going to heal his wounds.

I tossed and turned, but at some point I fell asleep. I guess I was more exhausted than I had realized, because the next time I opened my eyes it was morning. I reached over for Leon, but he wasn't there. I couldn't believe he had stayed downstairs the entire night. *I guess he no longer wants to sleep with me.* I rolled out of bed, washed my face and went downstairs to look for Leon. I checked the kitchen, the den, the living room, and the downstairs bathroom, but he was nowhere to be found. That's when I checked the garage and noticed his truck was gone. I ran upstairs to check the kids' rooms, hoping not to find them there. I was silently praying that because I had overslept, Leon had taken the kids to school. But they were still nestled in their beds. The telephone started ringing. I quickly ran to the bedroom to grab it thinking it had to be Leon calling.

"Hello."

"Sheila, did you decide to take the day off?" Tamara asked.

"Tamara, Leon's gone!"

"Huh? When? What happened?"

"Sometime between last night and this morning. When I woke up he was gone."

"Did you two argue last night?"

"No, he was very quiet. The last thing he said to me was that he needed some time alone."

"No note or nothing?"

"None that I see." I guess the phone had woke up Indya, she was walking into my room.

"Mommy, is it a holiday?" She asked me.

"No, honey. Mommy overslept."

"Where's daddy?"

"He had to run out."

"Is he coming back?"

"Of course, sweetie. Now go wash your face and brush your teeth. I'll have your breakfast ready in a minute." *Why in the heck did she ask me that? I guess she is more aware than I gave her credit for.*

"Mommy, you want me to wake up Kelsey?"

"No, just go do what I asked you please," I had started to lose my patience with her.

"Where do you think he is?" Tamara asked.

"I have no idea," I answered exasperated.

"Well, don't worry about coming in today. You need to be home when he gets back. In the meantime, I'll call Samuel to see if he has heard from him."

"Thanks, Tamara."

"Keep your head up, okay."

"I'll try. But I feel I'm slowly having a nervous breakdown."

"Don't even claim that. We will find him. He probably just needed to clear his head. The man had a lot to swallow on yesterday."

"And he appeared to be so unruffled by it all. It is like the calm before the storm and that's what frightens me."

"You don't think Leon will do anything to harm you, do you?"

"He does have a bit of temper and can get pretty crazy when he's mad."

"Then you go take the kids to school and I'll be by there later. If I hear from Samuel before then, I'll give you a call."

"Okay." I concealed my anxiety from Tamara. But I was praying that she would make it here before Leon did. His reserved demeanor had me concerned. You never know when a person is on the edge of insanity or what might make them snap. *Having your wife's lover call to tell you explicit details of their affair would be a good reason to make anyone lose it. So Leon not being here could just as well be a blessing in disguise.*

Chapter Forty-Four
Tamara

After I had hung up from speaking with Sheila, I quickly phoned Samuel to fill him in on the latest of Sheila and Leon's saga. He was on his way in to work, but he immediately changed directions and headed for Leon's job. Even though Wednesdays were Leon's regular day off, Samuel was hoping he went in to get his mind off things.

That's the difference between men and women. When faced with a problem men escape in their work. Women on the other hand, will eat, shop or cry and sometimes all at the same time.

I tied up some loose ends at work and was on my way out the door, but there was one more thing I needed to take care of before I left for the day and this just couldn't wait another second. It should have been done along time ago.

"Hi Haley, is he in?" I asked without slowing my stride.

"Yes, he is. But let me buzz him."

"No need," I continued walking until I reached Clyde's door. I didn't even bother to knock. He looked shocked to see me, or maybe it was fear.

"Tamara, what the…?"

"You low-down, no-good, trifling, phony minister of the gospel. How dare you try to sabotage Sheila's marriage? It wasn't bad enough that you seduced her when she was hurting and in need, but you had to call her husband and tell him all about it. You will pay for this. You're going down Reverend Peterson." I turned to walk out of Clyde's office and noticed someone was standing in the doorway.

"Tamara, this is my battle," Leon said with fire in his eyes.

I was shocked to see him and also scared at what he might do. *Where is Samuel when I need him? There is no way I could keep Leon off Clyde on my own.*

"Leon, let's just go. He's not even worth it."

"I'm not leaving until I get what I came for. I want Reverend Peterson to look me in my face and tell me why. Why my wife? I supported you. I trusted you. I worshiped with you. But never in my wildest dreams did I think I would have to worry about the pastor sleeping with my wife. Why?"

Leon was now standing directly in front of Clyde and I was getting nervous. The tension was thick. We were both waiting for Clyde to answer Leon's question. But before he could address it, security was at the door. Thank God, Haley had sense enough to call them.

They escorted Leon out the building and out to the parking lot to his car with me following closely behind. They wouldn't leave until he had gotten into his car and left the premises. I didn't ever get the chance to talk with him. I had no idea if he was heading home or not, so I called Samuel and told him to meet me at Sheila's house. Then I called Sheila to fill her in. Hopefully, Leon would be there when we arrived.

I made it to the house before either of them. Neither Samuel nor Leon's car were in sight.

"Hey girl. Have you heard from him?" I asked Sheila as I entered the house. I gave her a hug.

"No, he hasn't called and he's not answering his cell phone."

"Now, that he has had the opportunity to confront Clyde what do you think his next move will be?"

"I have no idea. I don't know where his head is. How was it that you happened to be in Clyde's office at the same time Leon was?" Sheila asked.

"Well, I have a confession to make. I had gone to Clyde's office to give him a piece of my mind. I'm sorry Sheila, but I couldn't hold it back any longer."

"I'm not even mad at you, girlfriend. It has taken everything inside of me not to contact him and let him have it. The only reason why I haven't is because I didn't want to make things worse than they already are. I have thought of everything from calling the church, to calling his supervisor at work. But I stop myself and realize we all reap what we sow. God will deal with Clyde in due time. He will not stand back and allow His sheep to be lead astray."

Samuel and Leon walked into in the house together. Leon went over to Sheila and gave her a hug.

"I'm sorry I made you worry," he said.

"We are going to get out of here and let you two be alone. Just call us if you need us," Samuel said. As he talked, he was leading me to the door. When we got outside, he gave me the biggest squeeze. "Boy, have I missed you," he said.

"I missed you, too and I have something wonderful to share with you."

"After these past few days I could stand to hear something wonderful."

"I called my mom last night!"

"Tamara, that's fantastic," he said kissing me.

"She was thrilled to hear from me and has kept up with my life through Aunt Josephine. She said she was very proud of me and my accomplishments and can't wait to meet you."

"Now that's great news. See, I knew things would work out and all it took was a phone call. I'm proud of you for taking the first step, baby. I love you."

"I love you, too."

As we drove off, going our separate ways, I thanked God for Samuel and for allowing us to meet. I vowed to do whatever it took to keep our relationship healthy and strong. After witnessing all the grief that Sheila and Leon were presently going through nothing or no one is worth coming between what we shared, not even my past.

Chapter Forty-Five
Leon

"I went to your job this morning to ask Reverend Peterson a few questions."

"I know. Tamara told me."

"I'm half glad that she was already there, because to tell you the truth, I'm not sure what else I would have said or done given the chance."

"I can't say I'm sorry enough, Leon."

"I know you're sorry, but why? Why did you cheat on me? I need to know."

"I didn't set out to cheat on you. It just happened."

"How in the hell do you just happen to sleep with another man? That's really cheap, Sheila!"

"I mean… I originally went to him for an advice and some guidance.

"Advice about what?"

"Us. Things had changed between us. I felt we were growing apart and I couldn't reach you. I went to him for help, but he manipulated the situation."

"You let him! You could have stopped it at anytime. So, don't you dare try to blame this on him. You had the power to say no, but you didn't. Why? I want to know what made you go through with it. What made you allow him to touch you?

"Because… I was lonely."

"How could you be lonely when I was right here every night? I wasn't running the streets or hanging out with the guys! I was here with my family! So you are going to have to do better than that, Sheila!"

"You were here in the house physically, but not emotionally. We were together, but we were only going through the motions. We stopped talking *to* each other and started talking *at* each other. We stopped dreaming together, which resulted in us leading separate lives and somewhere along the way we stopped caring."

"I never stopped caring! I'm not the one who was unfaithful! So you speak for yourself!"

"Leon, I'm…"

"Sorry. I know. But that's not helping me."

"What else can I do? I'll do anything Leon to get us back to where we were before this."

"Well, according to you, us being where we were was what got us where we are now. So maybe I'm just not the man for you."

"Leon, you're the only man for me."

"Apparently not! Do you realize what you did? You keep saying you're sorry, but what are you really sorry about? Are you sorry you got caught? Sorry that I found out? Or maybe sorry that I'm not him?"

"Leon, no! I'm sorry that I hurt you. I'm sorry that I allowed myself to be deceived."

"I'm sorry, too. Because I don't know if I can do this. How can I stay married to a woman who I have lost all faith in? I don't know who you are anymore. You were my everything. But now every time I close my eyes all I do is imagine you with that fraud Reverend Peterson. I see him pleasing you the way I used to and you enjoying it. Him knowing parts of you that only I knew. Just the thought of it makes me sick to my stomach! Everything that was dear to me has been shattered. I feel paralyzed. I can't get beyond the hurt." Leon paused to catch his breath. "I don't know where to go from here or how to get past it. How is it going to look when everyone finds out what you've done? How am I going to face my friends after this? I hope you don't except me to continue to go to that swindler's church! I will never set foot inside there again—and you either, if you know what's good for you! I love you Sheila, but you've hurt me more than you can ever know and I don't know if I can ever forgive you for this."

Chapter Forty-Six
Tamara

I met Samuel at another one of my favorite restaurants, Mikata's. They prepared the best Japanese dishes in town and the restaurant featured authentic oriental décor.

Once we were seated and had ordered, Samuel commented on how he really liked the atmosphere of the restaurant. I knew he would.

"So, how was the rest of your day?" I asked.

"You will never believe what happened."

"What?"

"Melanie quit!"

I remained silent. Part of me was happy and wanted to run and jump all over the room, but I had to be sensitive to Samuel's feelings.

"What happened?" I finally asked.

"When I arrived to work, she had left me a note saying she was resigning for a better opportunity. Not even two weeks notice or anything."

"Did you know that she was looking elsewhere?"

"No, she never said a word. I detected some distance from her since our last conversation regarding the unlikelihood of it being more between us. But I was really fine with that and sensed no reason for alarm, because she was still doing her job. Do you know I had to call a temp service?"

"I'm sorry, baby. Someone will come through."

"I know. I just hate having to start over with someone new. Melanie was very good at what she did. She helped to keep the business flowing smoothly."

"The next person will be even better. You'll see."

"Yeah, well I hope so. This couldn't have happened at the worst possible moment. I don't really have the time to train another assistant. In my business time is money."

"Is this the same man that told me that everything happens for a reason? Melanie may have been a good employee, but she also had ulterior motives. She wanted you for more than an employer. It's probably good that she left, because you never know what problems could have cropped up from her not getting her way."

"I just hope she doesn't try to keep Jared away from me. I have really grown attached to him."

"I'm sure in time she will do what's right."

"Enough about me. Have you heard anything from Sheila since earlier today?"

"No and I don't know if that's good or bad."

"Let's pray it's a good thing and they're on the road to healing their marriage. It's going to be tough, but I admire Leon for trying. Most men would have walked away and never looked back after hearing his wife had cheated on him. I certainly can't imagine what he's going through, but I know it has to be painful."

"I was shocked when Sheila told me about her affair with Clyde. I never would have guessed she would cheat on Leon."

"What do you think prompted her to do it?"

"I know she had made some comments in the past about Leon not being attentive, but she hadn't mentioned anymore to me lately. She had started confiding in Clyde. I saw him for the snake he was and I tried to warn her about his conniving ways, but she was convinced that he was on the up and up."

"A man like that gives the ministry a bad name."

"He's not a man, he's a reptile."

"As bad as what he did is, he still needs our prayers."

"How can you pray for him after what he's done to Sheila and Leon, and Lord only knows, who else?"

"Because somewhere along the way, he has been lead astray and he's going to need God to guide him back to the right path again."

"You can pray for him, but I won't."

"We have to hate the sin, but love the sinner. He is under the influence of the devil and he needs his fellow sisters and brothers to stand in the gap for him."

Obviously God wasn't finished with me, yet, because I couldn't find it in my heart to pray for Clyde. He was a scumbag in my book, who needed to be tarred and feathered for all the world to see.

Chapter Forty-Seven
Sheila

I guess I should be glad Leon didn't leave me after finding out about my infidelity, but this sudden silent treatment was driving me crazy. He barely acknowledged me and responded only when I asked a question, even then he didn't look me in the face when speaking to me. I have given up on ideas on how to make things right between us again and I certainly don't want to do anything to push him further away. I could forget about attempting any physical contact. He definitely wasn't having that. There was a time when I hoped he wouldn't touch me; now a slap on the butt would be a welcome.

I took the rest of the week off from work. This was done for the most part, because I didn't want to face Clyde. As strange as it may sound I was hurt by his actions. Not that I was under the impression that we were madly in love, but I never thought he would betray me the way he did either. I guess, Leon could say the same thing about me. If only I could turn back the hands of time, I would have never been unfaithful to Leon and be facing the reality of losing his trust in me. Him believing in me meant more than anything. It's a terrible feeling not to know the outcome of your future. But it also made it quite evident who truly was in charge. Hindsight is always 20/20.

Sitting at home alone I had plenty of time to reflect on all of the mistakes I had made in life. I went as far back as my first boyfriend in high school. He was the most popular boy in school and captain of the football team. I was the dorky smart girl, so I was flattered that he paid me any attention at all. I felt inferior to him and started to change who I was just to fit in. I changed the way I dressed, I changed the way I wore my hair and ultimately I changed me. In time that wasn't enough to keep Mr. Popularity, because he dumped me. That sent my sense of worth spiraling downward fast. Afterwards, I choose one bad guy after the other and my self-esteem continued to drop with each break-up.

By the time Leon came along I was a mess. I was hard on him and made him jump through a lot of hoops. Now I realize I was making him pay for all the past defective relationships I had been in before. It is amazing what a little time alone can do for the soul.

Before I could expect my marriage to improve, I had to chase these demons from my past out. I had to let go of all the anguish and resentment I had been holding on to for all these years. I had to release Leon from compensating me for what he had nothing to do with in the first place. It's painful to look deep inside yourself, because you are going to catch a glimpse of some things that you don't like. *I haven't realized just how wounded I was until now. All of this time I have been trying to change Leon, when the change should have taken place in me.*

I have also come to grips with who God, in fact, truly is. He was the creator of the universe, the guiding light, which means nothing happens without His say-so. Who was I to think I could control my own life without any instructions from Him? How foolish of me to think I could handle my marriage on my own terms. I didn't even hold the power to take my next breath. God controlled that too. Coming to the realization that I had no authority over my future or the outcome was very humbling. I relinquished all the control I thought I had at that moment. That was the first step towards my deliverance.

With my mind and spirit open to the Lord, His message to me was very clear. Leave. Uproot. Start over. A change was about to come.

Chapter Forty-Eight
Tamara

It had been a rough few days at work without Sheila. She took the remainder of the week off and her presence was definitely missed. I had to call her at home a couple of times to ask where she kept certain documents. It was becoming evident who were the real brains behind the operation. I was lost without her.

With everything that had been going on I still haven't had the chance to tell Sheila the news regarding my mom or the fact that she was planning a visit to come help me with some bridal preparations. The three of us were going to have a blast. Everything was falling into place and happening as it should in my life.

It was finally the weekend and my opportunity to spend some quality time with Samuel. He was still stressing about having to hire a new assistant. So, tonight we were going to spend a nice quiet evening at my place. I had rented a movie and he was bringing the food.

I had just enough time to spray the house with Febreze and change my clothes before Samuel arrived. He stepped through the door wearing khaki shorts, and a crisp white shirt holding takeout from Steak-out. Another one of my favorites.

"My hero. I'm starving," I said to Samuel as I reached for the bags.

"I aim to please."

"What would you like to drink?"

"Do you have any Coke?"

"Of course. I stopped by the video store and picked us up a movie, a comedy. I thought after the week we both had we needed a good laugh."

"You're right about that."

"Any luck with finding a new assistant?"

"No. I had to let the girl go that was sent over today."

"Why?"

"Let me just say her last job was at Micky D's. Nothing against the fast food industry, but I need someone with a professional demeanor. She had a gold tooth and her fingernails were six inches long."

I burst out laughing. "Well, I can sympathize with you. With Sheila being out this week, work has been horrible too. I'll be more than happy to see her come Monday morning. I may even throw her a party."

"Maybe you should. Good help is hard to find."

"I've been more worried about Sheila and Clyde's interaction with each other when she does return."

"I can understand that. It may not be conducive to a good work environment. Do their jobs require a large amount of contact with one another?"

"Not much, but enough to make things uncomfortable."

"Well, that's for Sheila to worry about, not us. I don't know about you, but I already have enough on my mind."

"Yeah, you're right and tonight is about you and me." I gave him a kiss and we stayed there for a few moments. I started to get excited and my womanly urges took over, but I quickly stopped myself. The consequences of having my way instead of doing things God's way flashed through my head. I lived that life for way too long. I would rather sacrifice now to receive a much bigger blessing down the road.

"Take your shirt off and lay down," I said to Samuel. He looked bewildered. "I want to give you a massage."

"What a treat. What have I done to deserve this?"

"Just by being you."

While Samuel took his shirt off and lay back as I had instructed, I went to my bedroom in search of some body oil. As I rubbed and gently kneaded his tired muscles, he said something that made me immediately stop the massage session.

"Maybe Sheila shouldn't come back."

Chapter Forty-Nine
Sheila

I'm not sure how much longer I could dance around the issue, but I needed to let Tamara know my future plans regarding my employment. I had stayed out yet another week. I know in my heart what I was going to do. It was the fact of telling Tamara that bothered me. I didn't want her to think that I was abandoning her, and I didn't want my decision to affect our friendship. Tamara has been very good to me and was much more than my supervisor. She was family.

Leon still wasn't talking to me and was spending less and less time at home. I guess his being distant and aloof was a part of my punishment for my sins. This was his way of making me pay. He spent time with the kids and continued relating with them, so it was apparent it was just me he was ignoring. The heaviness of it all was causing me to sink in a pit of depression. I knew I needed to snap out of it, but I couldn't. I prayed for an intervention.

Leon had already left for work (or wherever it was he was going these days). I had gotten the kids off to school and was preparing to crawl back into bed when the doorbell rang. I wasn't expecting anyone, so I snuggled deeper under the covers. The doorbell rang again and again. Whoever it was wasn't going away. So, I grabbed my robe and reluctantly went to answer the door. I opened the door to welcome Mama Lyn into the house and she came bearing gifts.

"Good morning, baby. I brought you by some breakfast," Mama Lyn said walking into the kitchen.

"Thank you, Mama Lyn, but I don't have much of an appetite."

"I don't want to hear nothing about your appetite. You need to eat. Now come, sit down and let me fix you something," she said as she stood in the kitchen and emptied the contents of her shopping bag. "I said sit down. I'm not gonna say it again."

"Yes, ma'am." I knew there was no use arguing with her.

"You look a mess," Mama Lyn commented, placing a plate of food in front of me.

"Thanks for noticing."

"Don't get sassy. I'm only telling you 'cause I love you. Now, I know you feel like your world has been turned upside down, but that is no excuse to let yourself go. You hiding away in this house ain't helping none either."

"Well, I don't know what else to do. Everything I've done in the past hasn't worked."

"Some of what you've done in the past didn't work, but prayer still does."

"Mama Lyn, I have prayed until I am sick of hearing myself. It's all on God now. I'm out of it."

"Well, I have news for you. God is not going to help you until you help yourself. You made a mistake, you've repented for it and now it is time for you to stop waddling in it. How are you and Leon getting along?"

"We aren't. He isn't here half the time and when he is, he acts as though I'm invisible."

"And that's exactly my point. Honey, you have got to get yourself together. The pity party is over. You need to resurrect the old Sheila, so that she can get her husband back before he strays too far."

"How am I going to that?"

"By doing what you did to get him in the first place, be Sheila. Don't make the mistake of trying to be his mother, his counselor, or his conscience. Men don't want another mother. They want a woman to love them unconditionally and to see them as their king. Never stop making him feel like he is the most important person in the world, and always make him feel as though he is in charge, even when he's not. Men need to feel needed. Leon is feeling that he has been replaced, so you have to reassure him that he still holds his position as man of the house. Do whatever it takes, but do something."

After Mama Lyn left, I took a long hot shower, fixed my hair, and made myself presentable. I cooked dinner for my family—meatloaf, Leon's favorite. I also called Tamara and told her that I wasn't coming back. She told me that she was going to miss me, but she understood. After all, family comes first. Now, all I have to do was tell Leon and hoped that he would be happy about it. I would find out soon enough.

"So, how was work today, honey?" I asked Leon.

"Same ole, same ole," he answered.

"How about you kids? How was school?" I asked.

"Same ole, same ole," Kelsey replied.

"Mine was good. It was a boy's birthday in my class and we had cupcakes!" Indya beamed.

"What kind of cupcakes did you have, sweetheart?" Leon directed his question to Indya.

"Chocolate! I ate two!"

"Well, it's good to know that it didn't spoil your appetite for dinner," Leon commented.

"Oh no, mom's meatloaf is the best," Indya said with a mouth full of food.

"Indya, no talking with your mouth full," I said to her.

"Sorry," Indya said stuffing her mouth with potatoes.

"If both of you are finished, you can be excused to do your homework," I told them.

"I don't have any homework," Kelsey answered.

"Well, go read a book," I said. I wanted some alone time with Leon.

"Aw man," Kelsey said as he got up from the table.

"I quit my job today," I blurted out as soon as the kids had left the room. But I received no response from Leon. He just looked at me. *At least I had his attention.* "I thought it would be best to completely start over and begin anew. A fresh beginning."

"You thought it would be good for who?"

"For us."

"You sure have been making a lot of decisions for us lately. Do I have any say-so in anything anymore or don't my opinion matter?"

"Of course, your opinion matters. You matter. That's why I decided to leave. I realized I made a bad choice before regarding our marriage, but I also realized that besides the kids you are the most important person in my life. Without you none of this works. You are my everything and I'm sorry that I failed to recognize that. But I can assure you that it will never ever happen again baby and I am willing to do whatever it takes for how ever long it takes to regain your trust in me again. I'll quit my job, I'll change my church membership, and I'll even leave town if you want us to. As long as we are together every step of the way, because, Leon, my life isn't worth living without you in it. You are my king and no one can ever take your place." I saw the first sign of a smile.

"Well, with that being said, you had better find yourself another job quick, because we need a check coming in here," Leon said. Then he reached over and held me with great tenderness.

Thank you God for Mama Lyn, my ram in the bush.

Chapter Fifty
Tamara

I had been walking around in a daze since Sheila turned in her resignation. I had a sick feeling in the pit of my stomach days before. It was as though it was an omen. I hated to lose her, but then again, I could definitely understand her motive. If only she had listened to me from jump street, we all would be happy now, all except Clyde. He wouldn't have gotten what he was after, and now that he had it appeared he was the only winner out of this. It didn't seem fair. If we all reaped what we sowed, I prayed that he would be reaping soon.

I was meeting Samuel for lunch. He still was having trouble finding the right assistant, but when I spoke with him earlier, he said he had found a solution and was going to let me in on it at lunch. I was open to anything, except Melanie returning.

The days were getting hotter, so the pantyhose that I was wearing were really giving me a fit. Who invented these anyway? Probably some man searching for perfection in women, because he was unable to find it in himself. When I entered the restaurant, Samuel's big grin was the first thing I spotted. Something had him in a good mood. I hope it was catchy.

"Hello, love. How is my wife-to-be?" Samuel asked as he stood up to greet me.

"Not as well as my future hubby. Tell me what has you all smiles."

"Let's order first. All this excitement has made me hungry." Samuel ordered the double club sandwich with a side order of fries and I ordered a chicken Caesar salad with a Diet Coke.

"Now, spill the beans."

"I want to hire Sheila as my new assistant."

"My Sheila?"

"Technically, she's not 'your Sheila' anymore. So, she's fair game. You've always talked about how proficient she was and with her spirit she would be just what I need in the office. But I wanted to make sure that you were okay with the idea."

I couldn't say anything. I knew it was selfish, but I wanted

Sheila for myself. But she did quit, and Samuel did need an assistant. Sometimes life can really suck.

"I guess if she isn't working for me, then you would be the next best choice. Go for it. I know you won't be sorry. Sheila is great at what she does. She would definitely be an asset. When are you going to call her for an interview?"

"An interview won't be necessary. You have already given her a high recommendation. I just need to know when she can start. There's something else."

"Don't tell me you want me to choose who I am going to have as a lunch buddy, you or her?"

"Not exactly. I want you to consider being my business partner. Before you answer just hear me out. The Lord has truly blessed my business. Clientele has doubled to the point where I am unable to maintain them on my own. I am in desperate need of a partner, someone I trust, someone who knows the business, and someone who provide excellent service to the clients. I prayed about it and came up with you. We would make an awesome team, and with Sheila holding things down at the office, we can't lose. I'll share everything with you straight down the middle and we could change the name to 'Roberts and Roberts Accounting Services'. How about it?"

"I think I like 'Roberts Accounting Firm' better."

"Yeah, that does have a nice ring to it. So, does that mean that you're coming aboard?"

"What about your not mixing business with pleasure saying?"

"Woman, your pleasure *is* my business. Seriously, my business is everything to me. I've nurtured it to what it is today. It took a lot of blood, sweat and tears to make it happen. It's my baby, and when I first envisioned it, I wanted it to be a legacy to be passed down through generations. Who else could I trust with the family business other than the woman I am going to share the rest of my life with?"

"Samuel, I don't know what to say."

"Well take your time in considering it. In the meantime, I'm gonna need Sheila's number."

I thought on Samuel's offer all day, but still had not made a concrete decision. That was until Clyde happened to be attending a meeting I was also in. The thought of working with him, and men like him, left a bad taste in my mouth. I was growing tired of corporate America and all the politics that went along with it. *Working alongside Samuel just may be the change I need.*

I smiled to myself as I thought of what the devil intended for bad, God turned into good. Because things were definitely working together for the good of us all who loved the Lord and who was called according to His purpose.

Chapter Fifty-One
Sheila

God is so good and is at work in my life. I was obedient and quit my job with Tamara and then another job opportunity was presented to me. Samuel called and offered me a job with his accounting firm! I was going to be working for a true man of God now. Leon would be pleased. I hope. I have already accepted the position, but I will talk it over with Leon as if I haven't. Just so he will feel a part of the decision making. I know what you're thinking. How deceitful! But whatever it takes to get us back on the right track I am willing to do.

Some more good news! Leon agreed for us to go to counseling. I had been thumbing through the Yellow Pages all morning trying to locate one. A Christian one. There were so many to choose from. But, at last, I ran across the perfect one for me and Leon. Crossroads Family Ministry. It was a husband and wife team. I called to inquire about their services. The receptionist was very warm and friendly. She informed me that they offered an initial consult for couples at no charge to ensure if it would be the right fit for them. I told her I would phone her back with a date in mind after speaking with my husband. I surely wasn't going to make the mistake of setting an appointment without Leon's input first.

I wanted so desperately for me and Leon to get back that loving feeling. You know the one where you can't think of anything else, but the one who holds your heart and brings a smile to your face at just the mention of their name. Not that I thought we didn't love one another, I think we had just stopped taking the time to truly express that love. I did loving things for Leon, but it was out of obligation; most of the times, it wasn't heartfelt. Through the years it had become more of a drudgery, than an expression of love. I considered it my duty as his wife to perform certain duties. So maybe it was my mindset that needed to be transformed. At any rate, counseling was definitely in order for us both.

I was also ready to get back into the work force. All of this time on my hands was slowly driving me insane. There are just so many loads of laundry you could wash in a day. I craved the stimulation of rushing into the day with options of the unknown, juggling my time between finishing a spreadsheet for Tamara and meeting payroll deadlines. I missed the interaction of my fellow co-workers and the girly chit chats we engaged in. I needed to get out of this house. My vacation was officially over!

After dinner I ushered the kids upstairs so they could get ready for bed. I needed to prepare Leon for my news bulletins. I wanted to guarantee that he was in the right frame of mind to receive what I had to report. I knew that he liked Samuel, but I didn't know if he would like me working for him. After everything that had transpired I expected it to be a touchy subject. But heck, I had to work somewhere.

"Would you like some help with the dishes?" Leon asked as he brought his plate to the sink.

"No sweetie. I got it. Besides, you have worked all day. Let me do something for you. How about a foot rub?"

"That's okay. I'll pass."

"A massage than. Full body," I said suggestively as I leaned in closer to him.

"Alright what's going on?"

"What are you talking about? I can't help my husband relax after a hard day on the job?"

"Yeah, but there is something behind it. I know you Sheila. So what is it?"

"I just need to talk to you. Give you an update on things."

"Things like what?"

"Like I found a counselor for us. It's Christian-based and run by a husband and wife. They offer an initial consultation at no price."

"Did you set up an appointment?"

"No, I wanted to check with you before I went ahead with anything."

"I don't have a problem with it, maybe we could try for Wednesday."

"Great. I'll call on tomorrow." We both just stood watching each other.

"What else?" Leon asked.

"I also received a job offer today."

"Well you're just full of surprises, aren't you? Where's the job?"

"I'd be working as an administrative assistant for Samuel. His last one quit and he asked me, seeing that I needed a job and all," I said it, all in one breath.

"Why did she quit?" Leon asked suspiciously.

"I didn't ask."

"Well maybe you should have. Inquiring minds would like to know," He said sarcastically.

"You know Samuel wouldn't have done anything out of the way to make her leave."

"No, I don't know that! After what happened with you and Reverend Peterson it has me questioning everything!"

"Leon, I know how you feel."

"You have no idea how I feel. I don't even know how I feel. Just do whatever, Sheila. I don't care." Leon turned to walk away.

"Leon don't do this. You said you wanted me to start including you in the decision making and that's what I'm doing. If you don't want me to take the job, I won't. But don't just walk away." I said to him as I grabbed him by the arm.

"Sheila, I'm confused. I don't know what to think, what to do or which way to turn. I look at everything and everybody differently. I don't know what to trust with my own eyes and I hate that. I just wish all of this had never happened and we could go back to how things were."

"I'm sorry."

"If I hear you say that you are sorry one more time I am going to hit something!"

"Well I am, Leon. If I could go back to undo things I would. But I can't. So I am trying the best I can to reconcile things. But I need you to help me."

"Did you tell Samuel that you would take the job?"

"No, not yet." *Okay I'm lying, but this is no time to be honest.*

"Do you want to?"

"Yes. I trust Samuel and his work ethnic. I think it's a great opportunity at just the right time. His business is really booming and I can help manage the growth and make a good contribution. But I'll only accept, if you are in agreement with me."

"Then take the job," Leon said as he walked out of the kitchen.

I was happy about the two decisions that would eventually better our future. But Leon's sour disposition was starting to worry me. We needed to begin counseling soon. *I am losing him little by little with each passing day.*

Chapter Fifty-Two
Tamara

Since my ace boon coon was no longer working with me and Samuel was busy on the job, I was forced to share my lunch with Cindy. I didn't exactly choose her; she actually chose me. I was sitting at a table in the cafeteria minding my own business when up she walked with that loud piercing voice of hers.

"Are you meeting anyone?"

"No," I answered.

"Good. So tell me why did Sheila leave?" Cindy asked before she could completely sit down. She wasn't wasting anytime.

"She was presented with a better opportunity." I tried to sound natural and casual, but I had to choose my words carefully because Cindy was known as a big gossip. I didn't want anything negative to come out about Sheila.

"Really. Where exactly?" she probed.

"With an accounting firm." I was beginning to become annoyed with all of her questions.

"So have you found someone to replace her, yet?"

I was shocked by the question and a little bit afraid of where it was going. I had to handle this with caution. "Why do you ask?"

But before she could answer Monique walked up.

"Hi ladies." She sat down without being asked, which was just like Monique.

"Join us please," I said scooting over to make room for her.

"So, how are you making it without Sheila?" Monique asked.

"Barely."

"Any prospects?"

"Not really," I answered, avoiding eye contact with Cindy.

"Well, I just heard that another position may be available soon because of a pending investigation," Monique said. She now had both of our attention.

Cindy literally stopped chewing and inched up closer to the table hanging on Monique's every word.

"Who?" I asked.

"Well, you didn't hear it from me, but it seems Clyde Peterson has been slapped with a sexual harassment suit."

"What?" I don't know if I had said that out loud or not, because my mind was buzzing with this startling news. I couldn't believe my ears, but to come from Monique it had to have some validity to it. She's not one to say something she doesn't know for sure was true. Now if it had slipped out of Cindy's mouth I would have just shrugged it off. She's a horse of a very different color.

"Who's the woman that is claiming sexual harassment?" Cindy asked and with a hint of an attitude, I might add. *What is her problem? Is she upset because Clyde didn't make a move on her? Or maybe she is upset because someone else shared the news before she could? You can never tell with her. She is a strange bird, to say the least.*

"The woman's name is not being disclosed at this time. Clyde hasn't been told who she is. But it is said she has irrevocable proof of the alleged accusation. Apparently, she had saved e-mails and voice messages from Clyde asking for sexual favors."

"Good girl," I commented.

"A good girl indeed," Monique said in her funniest Clyde-like voice. It was hilarious, except to Cindy, who had quietly slipped away probably to continue spreading the gossip. I couldn't wait to get to my office to call Sheila myself. This had made my day and I knew it would be a boost to Sheila's as well. We both wanted Clyde to pay for what he did to her. It looks as though that day is here. He finally rubbed up against the wrong woman, someone who wasn't taken in by his charm, position and good looks. Depending upon what was on those tapes, he wouldn't be able to wiggle his way out of this. *Oh yes, this is a good day indeed. Because Clyde's chickens are finally coming home to roost!*

Chapter Fifty-Three
Sheila

Leon and I had attended two counseling sessions and already our perspective on marriage had changed for the better. The two therapists were an older married couple who had been together for over 35 years. The man, Reverend Moss, was a retired pastor. He was tall and lean with a tranquil persona. Rose, his doting wife, was very small in stature. But what she lacked in height she made up in personality. We took to both of them from the very start.

The first thing they emphasized was that God created marriage as an example of His love. Our main objective in marriage is to please Him instead of ourselves. This brings to light the scripture, "Love the Lord your God with all your heart and with all your soul and with all your mind. This is the first and greatest commandment. And the second is like it: love your neighbor as yourself." I never thought of my husband as being my neighbor! My sole purpose was to serve Leon. That I did. But I was doing all the right things for all the wrong reasons. Our marriage was a ministry for others to see and witness. It wasn't about me getting my way all the time!

He also had us to pray for one another, not just in the mornings or before bedtime, but during the course of the day. Whenever the mood hit us we were told to call the other person and pray for them and with them. It was a nice feeling that whenever and whatever was on Leon's mind he shared it with me in prayer. I was beginning to get the concept. *Marriage is about building a relationship each day. Respecting and reverencing each other as the person God created. We were fashioned to be a blessing to each other and not a burden. It was a part of God's ingenious plan for us to be different and to have dissimilar strengths and weaknesses, so that we would complement one another.* I was seeing Leon in a new light. Thank God for fresh revelations!

This was my last week being unemployed. I had yet to begin working with Samuel, but must admit I was incredibly excited about it. I was going to be performing the same duties for him that I did for Tamara. Organizing and scheduling were my strengths. Sadly, I would miss the talks Tamara and I indulged in.

Not to mention the extended lunches and shopping sprees we would treat ourselves to from time to time. Working with Samuel was definitely going to be a change—for the better I hoped. He appeared to be a nice guy, but you never know a person until you worked along side them. I pray that he was a morning person. Lord knows Tamara wasn't. That was the only drawback about working with her. Many mornings I had to wind her up to get her ready for the day.

I was in the laundry room doing my second load of laundry. Folding clothes was one of my least favorite chores. I would rather be scrubbing the toilet. But I was on a roll when the phone rang. I was happy for the interruption.

"Hello."

"Hey, girl!" Tamara said on the other line.

"Hey!" I squealed into the phone.

"Are you busy?"

"Naw girl, I was just doing the laundry and was needing a break."

"I feel as though I haven't talked to you in ages. I miss you like crazy."

"Same here. It doesn't feel right not seeing and talking to you everyday. Though I'm excited about working with Samuel, it's not going to be the same. He's definitely not you!"

"Well, I have some news to share with you that may make you feel better," Tamara said with excitement in her voice.

"Don't keep me waiting. Spill the beans!"

"Samuel asked me to come to work with him as an equal partner!"

"Get out! Did you tell him 'Yes'?"

"You bet I did. It was an offer I couldn't refuse. So we are back working together again!" Tamara said, practically screaming in the phone!

"Tamara, this is perfect. How did he come to ask you to work with him? I thought he was in need of an assistant, not a partner. This is big!"

"After he asked me how I would feel about you working with him, he sprung it on me. He said he had prayed about it and that I was the answer that the Lord had given him. I was a little hesitant in the beginning, but that was before all the signs were presented to me. I took it as the Lord sending me confirmation."

"Girl, I was just pouting about missing you and God had already worked it out."

"Sheila, He is so good and allowing all my dreams to come true that it is almost scary. I keep thinking I am going to wake up soon and all of this is going to be gone."

"The Bible says He will open up the windows of heaven and pour you down a blessing that you won't have room enough to receive. I would say your windows are wide-open girlfriend. So don't be scared. Just be thankful."

"That I am. He's better to me than I could ever be to myself. I could just kick myself for struggling all of those years trying to do things on my own when all I had to do was surrender it all to Him. I wish somebody would have told me how easier my life would have been with God in control."

"Somebody did: Samuel. God knew how to get your attention." We both chuckled.

"I guess you have a point. So, what's been going on with you and Leon?"

"We've started counseling," I answered.

"Good. How has it been going?"

"Phenomenal! We're going to a husband and wife team. He's a retired minister and definitely anointed for this. We have only gone for two sessions and I have already noticed results."

"How did you find them?"

"The Yellow Pages. Their ad just jumped out at me."

"I'm so happy for you. I pray the very best for you and Leon. You know you are my adopted family and I can't have ya'll splitting up."

"I can say without a shadow of doubt that we are in this marriage for the long haul. It took me almost losing Leon to truly appreciate him for the awesome man God created just for me."

"You said a mouthful just then, girl. Are you sitting down? I have something else to tell you that is going to blow your mind."

"Wait a minute. Let me take a seat. Okay, go ahead."

"Someone has charged Clyde with sexual harassment and he's being investigated."

"What? Who?"

"No one is saying who she is, but she supposedly has solid evidence."

"What type of evidence?"

"She saved e-mails and voice messages from Clyde."

"If no one is saying who this woman is, how do you know there is a woman?"

"It came from Monique."

"'Nuff said."

"Exactly. I'm about to burst. Clyde's day is finally here." There was nothing but silence. "Sheila?"

"Yes, I'm here."

"What's wrong? I thought you would be as excited as I was to hear the news."

"It's not that. But…what if they ask for other women to come forward? Am I obligated too? Because things are finally coming together for Leon and I and I can't see putting him through that."

"First of all, no one is asking you to come forward, because this is not about you. This is about Clyde and some other woman that he violated. This is her fight."

"Well, whoever she is, I admire her courage. It takes a lot of nerve to come after Clyde Peterson. Courage, I obviously didn't have."

Chapter Fifty-Four
Tamara

It was just like old times earlier today while talking to Sheila. It felt as though no time had passed between us. I was even more excited about working with her again. She and I, along with Samuel, were going to make an awesome team.

The days were getting hotter and longer. It was too hot to ride with the top down. The sun was relentless. Summertime was undeniably here. With both of our busy schedules, Samuel and I hadn't spent a lot of time together. He appeared somewhat distant. I assumed the increase of his hectic workload was the culprit. He had declined the past few dinner invitations. But, at his request, tonight we were meeting at Sugar & Spice. There was urgency in his voice.

I rushed home from work to shower and change, before meeting Samuel. I slipped into an orange wrap dress and matching sandals. I reapplied my make-up and made a mad dash for the door. I didn't want to keep my future husband waiting.

In my haste, Samuel still made it to the restaurant before me. He was still wearing his suit, black with grey pinstripes, and a grey silk tie. Very classy. But something about him was different.

"Hi, sweetie," I greeted him. He stood as I approached, kissed me on the cheek and pulled out my chair. "Bad day at the office?" I asked with concern.

"I've had several bad days," he answered with a long face.

"Do you want to talk about it?"

"Tamara, you know how much I love you, right?"

"Of course, Samuel." All of a sudden I started to feel sick. *Lord please don't let him be breaking things off with me!* "What's wrong?"

"I don't know where to start. Everything just happened all at once," Samuel said more to himself than to me.

"What happened all at once?" Now I was really starting to panic.

"Tamara, I love you," Samuel said with intensity that made a chill run through my body.

"Samuel, please say what you need to say." My voice was trembling.

"I have a child," he spurted out. His eyes met mine, and then he looked away.

"I don't understand."

Our waiter made his way to our table to take our order. "Could you please give us a minute?" I snapped at him. He turned and swiftly walked away. I didn't mean to be rude, but this was no time for politeness.

"You recall me telling you about Rachel and the fact that she had aborted our baby, then left town. Apparently, she had the child." He stopped talking to take sip of the drink that was in front of him.

"How do you know this?" I asked, trying to hide my fear and concern.

"She contacted me a week ago and told me."

"A week ago and you're just now telling me?"

"Please, don't be upset. I had to digest it myself." He nervously fumbled with his napkin. "I met with her and my attorney earlier today." Now I was really beginning to worry. His words hung in the air.

"And?"

"She is demanding back child support, which I have no problem with paying. But I am also asking for joint custody."

Samuel has a son. A son I am not the mother of. What does this mean for us? I wasn't sure I want to share him with anyone else, even if that someone is his child.

Chapter Fifty-Five
Leon

I never wanted Sheila so badly, but every time I think of touching her my mind drifts back to her and Clyde. I couldn't get the vision of the two of them being together in the most intimate way possible out of my head. It hurt my heart each time I thought of it. *My wife with another man.*

We had gone to a few counseling sessions and though in the beginning I was a bit skeptical, it was helping with the war going on inside my head. But today not even the suggested phone prayers to one another were working for me. I needed to talk my feelings out man-to-man. So I called on Al. We hadn't really had the chance to talk since the men's conference, so I updated him on what was happening in my life lately. He was shocked to say the least. We agreed to meet at Buffalo's Cafe after work.

Buffalo's was a sports bar and favorite watering hole for most men working within thirty miles of the place. Men came from the boardrooms, classrooms, courtrooms, fire houses, post offices, and mechanic shops to relax and unwind from their day's troubles before heading home. It was our spot. No women. No drama.

Al was already inside when I arrived. He was at the bar yakking it up with Ken and Norman. They were regulars and known for debating about everything including our current president, religion, money, and their favorite topic—women. Which I found amusing seeing that neither of them had one to speak of and any money that they had was spent on the large quantities of beer that they drank. They were basically full of hot air and just talked to hear their own voices. Al and I left them arguing about, what else, women, to find us a table in the back.

"If nothing else those guys are always good for a laugh," I commented.

"They're harmless. I can respect a man with an opinion," Al said bringing his glass to his mouth. Al wasn't a drinker, so I can bet that was a Sprite he was nursing. "So, tell me how you are doing?" He asked.

"As well as expected, considering the circumstances." I shook my head.

"That really doesn't tell me anything."

"Well some days I think we're going to be okay and then there are days I don't want to go home from work. Guess what day today is?"

Al smiled. "I would think what you're feeling is normal. But you need to make up your mind which direction you're going to constantly go in. This fluctuation is not going to be good for you or your marriage. The fact that you do go home on the days that you don't feel like it tells me a lot. You still want your marriage to work. Am I right?"

"Of course, I still want it. It's just difficult. My heart wants to love Sheila and spoil her with everything imaginable. But then my mind takes over and says forget about what she wants. I need to take care of myself."

"But where would that get you? It seems a bit self-defeating and not the best way to help build on your relationship with Sheila. Your best defense would be to get your mind in line with your heart. Start today by forgetting what is behind you. When negative thoughts flood your mind, think positive. Focus on your future instead of digging around in the past. Let the dead stay exactly where it needs to be…buried."

"I'm trying, man."

"I know it's got to be hard, but you have to fight. The devil is after your marriage. That tells me there's a blessing on the other side of this. Don't give up when your breakthrough is right around the corner. Keep at it, and for goodness sake go home and make love to your wife. You look like crap."

We shared a laugh, but things got serious again.

"I love her so much, Al."

"So why are you still sitting here with me? Go home and remind Sheila of that."

Al was my mentor, someone I looked up to. I could always count on him to be nothing but brutally honest. He's not a man to beat around the bush. He says what he has to say and let's the chips fall where they may. That's much of the reason why I respect his opinion. I like the fact that he is not going to sugarcoat the situation, but he also will never steer me wrong. Al is like a father who chastises his son out of love. I knew whatever he said to me he meant it for my good.

Chapter Fifty-Six
Samuel

I have a son. A mini me. A little boy who needed me to guide him into manhood. I have a child to pass down generations of knowledge, wisdom and advice that I had learned from my father. It is going to be a challenge, but one that I am ready for. I lived for this moment most of my adult life. Cultivating creation.

The day Rachel walked into my office changed my existence forever. When I looked up from my desk and laid eyes on her, I thought my mind was playing tricks on me. Though it had been three years since the last time we had seen each other she still looked as I had remembered. I was so spellbound by seeing her that I didn't notice the toddler by her side.

Though she looked the same, something in her eyes appeared different. There were no pleasantries exchanged, only information. Within a few short minutes she conveyed to me that she had left town with the intent of aborting our child. But instead she accepted the life growing inside her and it changed her decision. She was determined to raise the child on her own. She choose not to inform me of her change of heart, because she knew my next move would be to urge her into marriage. She explained that she felt my only interest in marrying her was would have been because of her unexpected pregnancy, and she never wanted to be someone's wife out of obligation.

Long story short, life had not been kind to her. She was never able to fully rebuild her bookstore business in another location, which resulted in her having to file bankruptcy. She had to rely on government assistance to help her and Caleb, (that was my son's name), keep afloat. Now she was looking to get paid. She wanted money for all the years that she had been raising Caleb on her own. She asked nothing of my life or for my input regarding the matter. She wasn't interested. Her sole purpose was in meeting with me was to increase her bank account.

I knew it was going to be a bitter pill for Tamara to swallow. I, myself, was still trying to digest it. But I had to tell her. She had a right to know, because this was ultimately going to change her life as well as mine.

For several days I had avoided Tamara. The time had come for me to let the cat out of the bag. I thought meeting in a public place would be best. Maybe she wouldn't slap me if she had an audience. I struggled with opening up to her. Tamara had become such an essential part of my life and I didn't want to lose her. Not after I had finally found my other half—my soulmate.

I finally found the words to tell Tamara and awaited her response. The good news was she didn't slap me or run out the restaurant. She was rather calm, probably in shock. She seemed a bit miffed that I had just decided to tell her. What I tried to explain to her was that I was scared. I was afraid of finding one piece of who I was only to lose on another end. I wanted—no I needed—them both. *I pray Tamara was feeling the same about us, Caleb and me. Because I intend to fight for my son.*

Chapter Fifty-Seven
Tamara

I still hadn't convinced myself whether I was ready to take on the role of mother, to a child to whom I had not given birth. Would I be able to love him as my own? Would he accept me as his surrogate mother? This was very scary territory.

I was on my way to the cafeteria, lost in my thoughts when events happening before my eyes zapped me back to the moment. Clyde was being escorted out the building by security guards. One of them was carrying a box. Clyde was carrying his briefcase and had a grim look on his face. I knew he had to see me, but he appeared to look past me. No sexual innuendos, no flirtatious flattery, not even a glance. I stopped in my tracks and watched as they lead Clyde off the premises.

I continued on to the cafeteria hoping to run into somebody, anybody, who could tell me what was going on. I spotted Monique. She noticed me, too, and a huge grin spread across her face. I couldn't seem to get to her fast enough. It was as though my legs were in quicksand.

"What's happening? I just witnessed Clyde being accompanied to his car." I tried to whisper.

"His luck just ran out," Monique answered still smiling. "He's been terminated."

My mouth flew open. Monique continued to expound.

"It seems his unknown accuser had enough evidence for the charges to stick and he was asked to leave."

This was a powerful blow. Clyde, the king of charisma, was unable to talk his way out of this one. What was to come of his career, his marriage, and his reputation as a minister? Suddenly everything became crystal clear. My future with Samuel and the chance to partner with him in raising his son was a no-brainer. I was being given a once in a lifetime opportunity to marry the man of my dreams and I was actually considering letting him slip through my fingers because he came with a past. A past he was willing to make amends for where most men would shirk the obligation. God had blessed me with an extraordinary man and I couldn't wait to spend the rest of my life with him. I also couldn't wait to talk to him. I opted to skip lunch to go back to my office and telephone Samuel. All of a sudden there was urgency for me to hear his voice. I dialed his cell phone.

"Hey beautiful!"

"Samuel, I wanted you to know that I love you and even though I don't know Caleb, I love him, too. I can't wait until we're all one big happy family."

"Tamara, you just don't know how hearing you say that has made my day. I wasn't sure where you stood after hearing about Caleb. I know it was a shock. It was a shock to me too."

"I was just afraid of the unknown."

"I'm afraid, too baby. Being a father is unfamiliar territory for me, but I couldn't think of anyone I would want to experience it with other than you. Thanks for sticking it out with me," he said humbly.

"I wouldn't have it any other way."

"See ya tonight?"

"You're going to be seeing a whole lot of me soon. I'm typing up my resignation as we speak effective immediately. After today, I'm out of here. I don't care if this place burns down."

"Hallelujah! So what time will you be reporting to work in the morning?"

"In the morning! Who said anything about me starting work in the morning? I have shopping to do. I want to buy some new outfits so I can look good for my new partner."

"Baby, I don't think you are capable of looking any better than you do now."

"Now that comment just got you a body massage."

"I'm ready when you are."

"I'll see you around seven o'clock tonight at your place."

"Don't forget your oil."

"I never leave home without it."

After our conversation, I immediately called Sheila. I had to tell her all about Clyde. The phone rang several times before she finally picked up.

"Um…hello."

"Sheila?"

"Yes."

'I'm sorry, did I wake you?"

"Um... No. I was in the middle of something."

"Oh, okay. Well, I need to tell you something right quick. You will never believe…"

"Tamara, I'm a little busy right now. I will have to call you back." After several attempts with fumbling with the phone she hung up. What was that all about? I've never known her to be that dedicated to her housework before or maybe it wasn't the housework that had her busy. I smiled to myself as I thought of what could possibly be demanding Sheila's attention. *Go ahead girl! I ain't mad at ya!*

Chapter Fifty-Eight
Sheila

Prayer certainly changes things! I don't know what got into Leon, but I ain't complaining. He came from out of nowhere and laid it on me just like old times. He apparently was missing me as much as I was missing him and it showed. Once we connected there was no stopping us. Afterwards we cried in each other's arms. It felt like a rebirth. A beginning to our new life.

Leon took the kids to get ice cream. While they were out, I returned Tamara's call. What she had to tell me had been on my mind since we hung up earlier. My curiosity was piqued. I tried her home phone, but didn't get an answer. That didn't stop me. I simply dialed her cell phone.

"Oh, now you have time for me," Tamara said into the phone.

"Only a few minutes. Leon and the kids went out for some ice cream. So talk fast."

"What were you doing earlier that had you all wrapped up, tied up and tangled up that you were too busy to talk to your girl?" Tamara teased.

"Something you won't get a taste of until after you tell Samuel 'I do'."

"Oh, you're wrong for that. Just for that you can forget about my news I was going to share with you. You can just read about it in the Macon Telegraph."

"Girl, quit playing and spill the beans!"

"Are you sure you have time?"

"Yes!"

"Well, how about today I watched in delight as Clyde was escorted out of the building," Tamara announced.

"What!"

"The harassment charges stuck and he was relieved of his duties."

"I would have loved to have been there just to see the look on his face."

"Honey, it was pitiful. He wasn't cracking a smile. Old Clyde didn't have anything to say today."

"I wonder what is going to happen when the church finds out," I said.

"Probably nothing. Those people had to know what type of man he was when they offered him the position. As long as he keeps them entertained and the money continues to flow, he'll have a job."

"I can't see God allowing His sheep to be lead astray for long.

"It will go on for as long as the sheep tolerates it. Most people don't want to change. They just want to go to church and made to feel good. They are not going to question anything unless it directly affects them. What do they care if he was charged with sexual harassment as long as the woman wasn't their wife? So don't count on those deacons to take a stand. I've been through it all before. The church is a strong force to try to come up against and you know how members love their pastor. They are ready to fight you if say something bad about him."

"Now, girl you ain't spoke nothing but the truth just then!"

"I also turned in my letter of resignation today."

"Good for you! Hey, I hear my crew coming through the door. I will give you a call back tomorrow."

"Please do. We have a lot more catching up to do and wedding plans to make."

"Most definitely. I love you, girl!"

"Love you, too. Tell the family I said hello."

I ran down the stairs and met everyone in the kitchen. I believed Leon had bought every flavor of ice cream Baskin Robbins sold. We laughed and talked as we tried them all. We each had our own bowl and dipped our spoons in each other's to taste a different flavor. *This is what life was all about. Family.* But I must admit I was ready for the kids to retire to their beds, so that Leon and I could continue where we had left off earlier. It was so good to have my old Leon back and I was determined to keep him around.

Chapter Fifty-Nine
Sheila

As I lay in the bed listening to Leon sing in the shower I reflected on our past year together. We had gone from being merely roommates to being the replica of love God had created us to be. After finishing our counseling sessions we gleaned such valuable information to assist us on our road to healing that we felt compelled to share it with other couples. We joined the couples' ministry at church and eventually began to head the meetings. With us working through the tough times, instead of heading for divorce court, it equipped us to help other couples during their time of conflict. Our test became our testimony.

It didn't take long for the news to get out about Clyde's dismissal from his job, because of sexual harassment, to reach the congregation. It turned out that Monique was the woman who filed the charges against him. Clyde tried his best to persuade the Deacon's Board that the accusations were all lies against him, but the evidence was too staggering, especially when a few ladies from the church spoke out about him as well. He was voted out and relieved of his duties immediately. Last I heard he had moved to Atlanta and was preaching out of what used to be an old crack house.

My affair with Clyde served as a lesson that looks can be deceiving and as Mama Lyn had told me 'Many are called, but few are chosen'. Just because Clyde called himself a pastor, didn't make him one.

Yes, Leon and I had a reason to rejoice. But the year also had its share of tears. I lost my Mama Lyn. She had contracted pneumonia and never recovered. But even in her last days Mama Lyn's faith in God never wavered. Ultimately, she praised her way through Heaven's gate. Her spirit was infectious and lives on inside of me today. So on days when I miss her so badly that I find it hard to breathe, I think of the days I had with her and the lessons she taught me. No, she didn't leave behind millions, but she did leave a wealth of knowledge.

Chapter Sixty
Tamara

My wedding day was right out of a fairy tale. All the people that I loved were there to watch me walk down the aisle to my Prince Charming. My mother and Aunt Josephine were on the front row. I wasn't able to contact my dad, though I tried. The number I had for him had been disconnected. I didn't even know if he was alive or dead. But I finally was able to free my heart of all the resentment I felt toward him for so many years. Life was too short to dwell on the past.

Caleb was our ring bearer, and a very handsome one at that. Samuel succeeded in gaining joint custody of him and we had bonded quite nicely. Now don't get me wrong, it was no piece of cake. I had to grow into suddenly becoming a mother months before me and Samuel were even married. There were days when dealing with Rachel and trying to connect with Caleb seemed too overwhelming. But I seen the love in Samuel's eyes that he had for him, it touched me deeply and I couldn't imagine walking away.

Now I was reaping the benefits for being obedient, because I was living the good life I had always dreamed of. Shortly, after I joined Samuel's business as partner our clientele appeared to have tripled and we had to open up a second location. He remained in the downtown location, while Sheila and I relocated to an office on Riverside Drive.

God had truly blessed us and gave us back what the devil tried to destroy. I am happier than I have ever been. That's not to say that I don't encounter difficult days; I simply know now just how to praise my way through.